T0326390

Farm animal proteomics 2014

Farm animal proteomics 2014

Proceedings of the 5th Management Committee Meeting and
4th Meeting of Working Groups 1,2 & 3 of COST Action FA 1002

Milano, Italy

17-18 November 2014

edited by:
André de Almeida
Fabrizio Ceciliani
David Eckersall
Ingrid Miller
Jenny Renaut
Paola Roncada
Romana Turk

Wageningen Academic
P u b l i s h e r s

EAN: 9789086862627
e-EAN: 9789086868100
ISBN: 978-90-8686-262-7
e-ISBN: 978-90-8686-810-0
DOI: 10.3920/978-90-8686-810-0

Cover drawing by Simão Mateus

First published, 2014

© Wageningen Academic Publishers
The Netherlands, 2014

The individual contributions in this
publication and any liabilities arising from
them remain the responsibility of the authors.

Neither the COST Office nor any person
acting on its behalf is responsible for the use
which might be made of the information
contained in this publication. The COST
Office is not responsible for the external
websites referred to in this publication.

The publisher is not responsible for possible
damages, which could be a result of content
derived from this publication.

The COST Organisation

COST – European Cooperation in Science and Technology – is an intergovernmental framework aimed at facilitating the collaboration and networking of scientists and researchers at European level. It was established in 1971 by 19 member countries and currently includes 35 member countries across Europe, and Israel as a cooperating state. COST funds pan-European, bottom-up networks of scientists and researchers across all science and technology fields. These networks, called 'COST Actions', promote international coordination of nationally-funded research.

By fostering the networking of researchers at an international level, COST enables break-through scientific developments leading to new concepts and products, thereby contributing to strengthening Europe's research and innovation capacities.

COST's mission focuses in particular on:
* building capacity by connecting high quality scientific communities throughout Europe and worldwide;
* providing networking opportunities for early career investigators;
* increasing the impact of research on policy makers, regulatory bodies and national decision makers as well as the private sector.

Through its inclusiveness, COST supports the integration of research communities, leverages national research investments and addresses issues of global relevance. Every year thousands of European scientists benefit from being involved in COST Actions, allowing the pooling of national research funding to achieve common goals. As a precursor of advanced multidisciplinary research, COST anticipates and complements the activities of EU Framework Programmes, constituting a 'bridge' towards the scientific communities of emerging countries. In particular, COST Actions are also open to participation by non-European scientists coming from neighbour countries (for example Albania, Algeria, Armenia, Azerbaijan, Belarus, Egypt, Georgia, Jordan, Lebanon, Libya, Moldova, Montenegro, Morocco, the Palestinian Authority, Russia, Syria, Tunisia and Ukraine) and from a number of international partner countries.

COST's budget for networking activities has traditionally been provided by successive EU RTD Framework Programmes. COST is currently executed by the European Science Foundation (ESF) through the COST Office on a mandate by the European Commission, and the framework is governed by a Committee of Senior Officials (CSO) representing all its 35 member countries. More information about COST is available at www.cost.eu.

This publication is supported by COST.

Editors

- André M. de Almeida, CVZ-FMV, Universidade de Lisboa, Av. Univ. Técnica, 1300-477 Lisboa, Portugal; aalmeida@fmv.utl.pt
- Fabrizio Ceciliani, Department of Veterinary Science and Public Health Università di Milano Via Celoria 10, 20133 Milano, Italy; fabrizio.ceciliani@unimi.it
- David Eckersall, Institute of Biodiversity, Animal Health and Comparative Medicine, R329 Level 3, Institute of III, Jarrett Building, Glasgow G61 1QH, United Kingdom; p.d.eckersall@vet.gla.ac.uk
- Ingrid Miller, Institute for Medical Biochemistry, University of Veterinary Medicine, Veterinärplatz 1, 1210 Vienna, Austria; ingrid.miller@vetmeduni.ac.at
- Jenny Renaut, Centre de Recherche Public - Gabriel Lippmann, Rue du Brill 41, 4422 Belvaux, Luxembourg; renaut@lippmann.lu
- Paola Roncada, Istituto Sperimentale Italiano Lazzaro Spallanzani, c/o Department of Veterinary Science and Public Health, Università di Milano, Via Celoria 10, 20133 Milano, Italy; paola.roncada@guest.unimi.it
- Romana Turk, Department of Pathophysiology, Faculty of Veterinary Medicine, University of Zagreb, Heinzelova 55, 10000 Zagreb, Croatia; rturk@vef.hr

Table of contents

The COST Organisation   5

Proteomics in farm animals: all roads leading to Milano   15

Part I – Oral communications

Perspective in comparative proteomics research   19
Andrea Urbani

Changes in the proteome of mastitis-causing *Escherichia coli* strains that affect pathogenesis 21
John Lippolis

Colostrum protein uptake in neonatal lambs examined by descriptive LC-MS/MS   22
Lorenzo E. Hernández-Castellano, Anastasio Argüello, André M. Almeida, Noemí Castro
and Emøke Bendixen

Proteomic investigation in milk whey from healthy and mastitis affected Jafarabadi
buffaloes (*Bubalus bubalis*)   26
André M. Santana, Daniela G. Silva, Funmilola Thomas, Richard J.S. Burchmore,
José J. Fagliari and Peter D. Eckersall

Bovine LDL and HDL proteome in health and inflammatory state   29
Milica Kovačević Filipović, Emily L. O'Reilly and P. David Eckersall

Application of top-down proteomics in detecting biomarkers in porcine saliva samples   33
María Fuentes-Rubio, Monica Sanna, Tiziana Cabras, Massimo Castagnola,
Federica Iavarone, F. Tecles, J. Ceron and Irene Messana

Proteomic characterization of molecular changes in saliva proteins in stressed pigs   37
Lourdes Criado, Laura Arroyo, Daniel Valent, Anna Bassols and Ingrid Miller

Identification of welfare biomarkers in pigs using peripheral blood mononuclear cells
and differential gel expression   41
Daniel Valent, Laura Arroyo, Lourdes Criado, Francesca Dell'Orco and Anna Bassols

A search for protein biomarkers links olfactory signal transduction to social immunity   45
M. Marta Guarna, Andony P. Melathopoulos, Elizabeth Huxter, Immacolata Iovinella,
Robert Parker, Nikolay Stoynov, Amy Tam, Kyung-Mee Moon, Queenie W. T. Chan,
Paolo Pelosi, Rick White, Stephen F. Pernal and Leonard J. Foster

Bacterial competition for food safety in dairy products 47
Isabella Alloggio, Mylène Boulay, Vincent Juillard, Véronique Monnet, Alessio Soggiu,
Cristian Piras, Luigi Bonizzi and Paola Roncada

Mule duck 'foie gras' show different metabolic states according to their quality
phenotypes by using a proteomic approach: comparison of 2 statistical methods 51
Yoannah François, Christel Marie-Etancelin, Alain Vignal, Didier Viala, Stéphane Davail
and Caroline Molette

Differential peptidomics of raw and pasteurized sheep milk cheese 55
M. Filippa Addis, Salvatore Pisanu, Daniela Pagnozzi, Massimo Pes, Antonio Pirisi,
Roberto Anedda and Sergio Uzzau

The Grana Padano microbiota: insights from experimental caseification 59
Alessio Soggiu, Cristian Piras, Stefano Levi Mortera, Milena Brasca, Andrea Urbani,
Luigi Bonizzi and Paola Roncada

Supplemented diets for allergenic modulation in European seabass 62
Denise Schrama, Jorge Dias and Pedro M. Rodrigues

Protein changes in muscles with varying meat tenderness 66
Eva Veiseth-Kent, Kristin Hollung, Vibeke Høst and Rune Rødbotten

Intact cell MALDI-TOF MS: a male chicken fertility phenotypic tool 72
Laura Soler, Aurore Thèlie, Valérie Labas, Ana-Paula Teixeira-Gomes, Isabelle Grasseau,
Grégoire Harichaux and Elisabeth Blesbois

Identification of novel protein interactions related to spermatid elongation 76
Mari Lehti, Noora Kotaja and Anu Sironen

Proteomic analysis of different bovine epithelial cells phenotypes after *Escherichia coli*
LPS challenge 80
Cristian Piras, Yongzhi Guo, Alessio Soggiu, Viviana Greco, Andrea Urbani, Luigi Bonizzi,
Patrice Humblot and Paola Roncada

Proteome adaptation of boar semen during storage 84
Viviana Greco, Blanka Premrov Bajuk, AlessioSoggiu, Cristian Piras, Petra Zrimšek,
Maja Zakošek Pipan, Luigi Bonizzi, Andrea Urbani and Paola Roncada

Metaproteomic analysis, application to water quality assessment in aquaculture 88
Jacob Kuruvilla, Binu Mathew, Gabriela Danielsson, Alessio Soggiu, Paola Roncada and
Susana Cristobal

Fat&MuscleDB: integrating 'omics' data from adipose tissue and muscle   91
Jérémy Tournayre, Isabelle Cassar-Malek, Matthieu Reichstadt, Brigitte Picard,
Nicolas Kaspric and Muriel Bonnet

Developing new methods for PTM mapping of *Ehrlichia ruminantium* proteins   95
Isabel Marcelino, Núria Colomé-Calls, Rita Laires, Thierry Lefrançois, Anna Bassols,
Nathalie Vachiéry, Ana V. Coelho and Francesc Canals

Integrative label-free quantitative proteomics study in mastitis   98
Manikhandan A.V. Mudaliar, Funmilola C. Thomas, Mark McLaughlin,
Richard Burchmore, Pawel Herzyk, P. David Eckersall and Ruth Zadoks

Proteomic profiling of the obligate intracellular bacterial pathogen *Ehrlichia*
ruminantium outer membrane fraction   102
Amal Moumène, Isabel Marcelino, Miguel Ventosa, Olivier Gros, Thierry Lefrançois,
Nathalie Vachiéry, Damien F. Meyer and Ana V. Coelho

NMR for metabolomic tissue profiling in small ruminants: a tool for proteomics study
complementation   103
Mariana Palma, Manolis Matzapetakis and André M. Almeida

Part II – Proteomics in infection diseases and animal production

Proteomics for assessing quality and metabolic compatibility of aquaculture feeds   109
M. Filippa Addis, Roberto Anedda, Grazia Biosa, Elia Bonaglini, Roberto Cappuccinelli,
Stefania Ghisaura, Riccardo Melis, Daniela Pagnozzi, Simona Spada, Hanno Slawski and
Sergio Uzzau

Finding novel milk protein markers for small ruminant mastitis   112
Maria Filippa Addis, Salvatore Pisanu, Vittorio Tedde, Stefania Ghisaura, Grazia Biosa,
Daniela Pagnozzi, Tiziana Cubeddu, Stefano Rocca, Gavino Marogna, Ignazio Ibba,
Marino Contu, Simone Dore, Agnese Cannas and Sergio Uzzau

Proteomic identification of immunogenic proteins in *Leishmania infantum*   115
Katarina Bhide, Carmen Aguilar-Jurado, Sara Zaldivar-Lopez, Ignacio López-Villalba,
Manuel Sánchez-Moreno, Ángela Moreno and Juan J. Garrido

PilE4 of *Francisella* alters expression of proteins on the brain endothelium   119
Mangesh Bhide, Elena Bencurova, Andrej Kovac, Lucia Pulzova and
Zuzana Flachbartova

Farm animal proteomics as a tool for understanding human health: the case of the
ACOS-innovation Project   123
*Luigino Calzetta, Mario Cazzola, Andrea Urbani, Paola Rogliani, Paola Roncada and
Luigi Bonizzi*

Proteomics of the mitochondrial proteome in dairy goats (*Capra hircus*)   128
*Graziano Cugno, Lorenzo E. Hernandez-Castellano, Mariana Carneiro, Noemí Castro,
Anastasio Argüello, Juan Capote,Sébastien Planchon, Jenny Renaut,
Alexandre M. Campos and André M. Almeida*

A proteomic study of bovine whey in a model of Gram positive and Gram negative
bacterial mastitis   131
*Funmilola C. Thomas, Timothy Geraghty, Patricia B.A. Simoes, Lorraine King,
Richard Burchmore and Peter D. Eckersall*

Blood haptoglobin concentration in New Zealand white rabbits during first year of their life   135
*Teodora M. Georgieva, Vladimir S. Petrov, Aanna Bassols, Evgenya V. Dishlyanova,
Ivan P. Georgiev, Kalina N. Nedeva, Mariela I. Koleva, Fabrizio Ceciliani and
Tatyana Vlaykova*

A comparative proteomic analysis of human umbilical vein endothelial cells after
infection with the rodent-borne hemorrhagic fevers puumala hantavirus and *Leptospira
interrogans* serovar Copenhageni   140
*Marco Goeijenbier, Byron E.E. Martina, Marga G.A. Goris, Ahmed Ahmed,
Rudy Hartskeerl, Sebastién Planchon, Kjell Sergeant, Jenny Renaut, Eric C.M. van Gorp,
Jarlath Nally and Simone Schuller*

Profile changes in salivary glyco-enriched fraction of pigs with inflammation   144
*A.M. Gutiérrez, I. Miller, M. Fuentes-Rubio, K. Hummel, K. Nöbauer, E. Razzazi-Fazeli
and J.J. Cerón*

A proteomics study on colostrum and milk proteins of the two major small ruminant
dairy breeds from the Canary Islands on a bovine comparison perspective   148
*Lorenzo E. Hernández-Castellano, André M. Almeida, Sébastien Planchon, Jenny Renaut,
Anastasio Argüello and Noemí Castro*

A 2DE map of the urine proteome in the cat: effect of Chronic Kidney Disease   154
*Enea Ferlizza, Alexandre Campos, Aurora Cuoghi, Elisa Bellei, Emanuela Monari,
Francesco Dondi, André M. Almeida and Gloria Isani*

Serum proteomic analysis of zoonotic-related abortion in cows 158
Matko Kardum, Cristian Piras, Alessio Soggiu, Viviana Greco, Paola Roncada,
Marko Samardžija, Silvio Špičić, Dražen Đuričić, Nina Poljičak Milas and Romana Turk

Comparison of quail and chicken broiler circulating IGF-1 162
Virge Karus, Avo Karus, Harald Tikk, Aleksander Lember and Mati Roasto

A proteomic profile of uncomplicated and complicated babesiosis in dogs 166
Josipa Kuleš, Carlos de Torre, Renata Barić Rafaj, Jelena Selanec, Vladimir Mrljak and
Jose J. Ceron

Study of the effects of saturated fatty acids on the porcine epitheloid ileum IPI-2I cell line 171
Anna Marco-Ramell, Kerry Wallace, Michael Welsh, Gordon Allan, Mark Mooney and
Violet Beattie

Gustducin gene expression in sheep, goats and water buffalo to unravel taste signaling 175
Andreia T. Marques, Ana M. Ferreira, Susana S. Araújo, Laura Restelli, Cristina Lecchi,
André M. Almeida and Fabrizio Ceciliani

Molecular studies on adipose tissue in turkey: searching for welfare biomarkers 179
Andreia T. Marques, Sara Rigamonti, Cristina Lecchi, Guido Grilli, Sara Rota Nodari,
Leonardo James Vinco and Fabrizio Ceciliani

Modelling of 3-oxoacyl syntheses 2 from *Brucella suis* to be used for structure based
drug design 184
Jani Mavromati, Dimitrios Vlachakis, Sophia Kossida and Xhelil Koleci

Proteomic analysis of gilthead sea bream plasma with amyloodiniosis 188
Márcio Moreira, Denise Schrama, Florbela Soares, Pedro Pousão-Ferreira and
Pedro Rodrigues

The measurement of chicken acute phase proteins using a quantitative proteomic approach 193
Emily L. O'Reilly, P. David Eckersall, Gabriel Mazzucchelli and Edwin De Pauw

Immunoproteomic analysis of *Mycoplasma meleagridis* proteins 197
Ticiana S. Rocha, Alessio Soggiù, Maurizio Ronci, Luigi Bertolotti, Paola Roncada,
Salvatore Catania, Andrea Urbani, Luigi Bonizzi and Sergio Rosati

Comparative analysis of acute phase proteins in farm animals by affinity methods
coupled to proteomics 200
Lourdes Soler, Fermín Lampreave, M.A. Álava, Richard J.S. Burchmore and
Peter D. Eckersall

An integrative study of the early immune response against ETEC   204
Laura Soler, Marcel Hulst, Jan van der Meulen, Gabriel Mazzucchelli, Mari Smits,
Edwin de Pauw and Theo Niewold

Effect of the use of OTC as feed additive in the pig serum proteome   207
Laura Soler, Ingrid Miller, Karin Hummel, Flemming Jessen, Manfred Gemeiner,
Ebrahim Razzazi-Fazeli and Theo Niewold

Total alkaline phosphatase and bone alkaline phosphatase in dairy cows during
periparturient period   210
Jože Starič, Marija Nemec and Jožica Ježek

Genomics and deep proteome profiling of *Staphylococcus epidermidis* for uncovering
adaptation and virulence mechanisms   214
Pekka Varmanen, Pia Siljamäki, Tuula A. Nyman and Kirsi Savijoki

Blood fibrinogen concentration in New Zealand White Rabbits during first three months
of their life   217
Evgenya V. Dishlyanova, Teodora M. Georgieva, Vladimir S. Petrov, Tatyana Vlaykova,
Fabrizio Ceciliani, Radina N. Vasileva and Ivan P. Georgiev

Part III – Proteomics analysis of food from animal origin

Using shotgun proteomics to understand the effect of feed restriction on the *Ovis aries*
wool proteome   223
André Martinho Almeida, Jeffrey E. Plowman, Duane P. Harland, Ancy Thomas,
Tanya Kilminster, Tim Scanlon, John Milton, Johan Greeff, Chris Oldham and
Stefan Clerens

Comparative proteomic analysis of muscle tissue from pre-term and term calves 226
Paula Friedrichs, Hassan Sadri, Julia Steinhoff-Wagner, Harald Hammon,
Allan Stensballe, Emøke Bendixen and Helga Sauerwein

High frequencies of the α_{S1}-casein zero variant and its relation to coagulation properties
in milk from Swedish dairy goats   230
Monika Johansson, Madeleine Högberg and Anders Andrén

First characterization of the goat mammary gland proteome secretory tissue using
shotgun proteomics 234
*Joana R. Lérias, Lorenzo E. Hernández-Castellano, Noemí Castro, Anastasio Argüello,
Juan Capote, Alan Stensballe, Jeffrey E. Plowman, Stefan Clerens, Emoke Bendixen and
André M. de Almeida*

Identification of potential biomarkers of animal stress in the muscle tissue of pigs caused
by different animal mixing strategies 239
*A. Rubio-González, M. Oliván, Y. Potes, D. Illán-Rodríguez, I. Vega-Naredo, V. Sierra,
B. Caballero, E. Fàbrega, A. Velarde, A. Dalmau, F. Díaz. and A. Coto-Montes*

Novelty and tradition: when proteomics meets Nero di Parma ham 240
*Gianluca Paredi, Samanta Raboni, Roberta Virgili, Alberto Sabbioni and
Andrea Mozzarelli*

Can zymographic analysis of proteases activities provide new informations on 'foie gras'
cooking losses? 243
Hervé Rémignon, Nathalie Marty-Gasset and Sahar Awde

Peptidomics as a robust and reliable approach to discriminate between closely-related
meat animal species 248
Alberto Massa, Enrique Sentandreu, Carlos Benito and Miguel A. Sentandreu

Proteomic analysis of adipose tissue from peripartum high yielding dairy cows 253
Maya Zachut

Part IV – Advancing methodology for farm animal proteomics

Detection of whey fraction common proteins of human and goat colostrum by MALDI-
TOF/TOF 259
Cansu Akin, Sébastien Planchon, Jenny Renaut, Ugur Sezerman and Aysel Ozpinar

Regional brain neurotransmitter levels and proteomic approach: sex, halothane genotype
and cognitive bias 261
*Laura Arroyo, Anna Marco-Ramell, Raquel Peña, Daniel Valent, Antonio Velarde,
Josefa Sabrià and Anna Bassols*

Computational study of interaction of borrelial ospa with its receptors 265
*Elena Bencurova, Dimitrios Vlachakis, Lucia Pulzova, Zuzana Flachbartova,
Sophia Kossida and Mangesh Bhide*

Recent advances in HRAM quantification: application to clinical assays   270
Bruno Domon

The sheep (*Ovis aries*) muscle proteome: decoding the mechanisms of tolerance to
Seasonal Weight Loss using label free proteomics   271
Ana M. Ferreira, Paolo Nanni, Tanya Kilminster, Tim Scanlon, John Milton, Johan Greeff,
Chris Oldham and André M. Almeida

Species determination of animal feed by QQQ mass spectrometry   275
Yue Tang, Janine Gielbert and Jim Hope

Differences between CH1641 scrapie and BSE prions identified by quantitative mass
spectrometry   277
Adriana (Janine) Gielbert, Yue Tang, Maurice J. Sauer and James Hope

Proteomics data from ruminants easily investigated using ProteINSIDE   283
Nicolas Kaspric, Brigitte Picard, Matthieu Reichstadt, Jérémy Tournayre and
Muriel Bonnet

Acknowledgements   289

Proteomics in farm animals: all roads leading to Milano

The COST Action FA on Farm Animal Proteomics has undoubtedly provided a major stimulus to the applications of advanced proteomics to studies in farm animal and aquaculture sciences during production and subsequent harvest. An international forum for development of applications and related technology has been created within the 4 years of the Action with collaborations across Europe greatly enhanced.

From our first Working Group Meetings in Glasgow (UK), through those in Vilamoura (Portugal), Kosice (Slovakia) to the Final Showcase Meeting in Milan (Italy) there have been an enthusiastic and enervating gatherings with significant cross-fertilisation of knowledge and ideas between experts on proteomic technology and those in the widest range of farm animal sciences from study of diseases of dairy cows, to species identification of fish and conversion of muscle into meat. Throughout it has been particularly rewarding that many early stage researchers have taken the opportunity provides by short term scientific missions to travel and explore the scientific potential of collaborating with laboratories outside their national boundaries.

It is imperative that the success of the COST Action in establishing a viable network of experts in farm animal proteomics is maintained and a means devised to ensure its continuation. There are multiple applications of proteomics and it is important that its potential to contribute and enhance the outcome of studies across the disciplines is recognised and supported by national and international funding agencies. In particular, it is essential that the full potential of proteomics is included in the programmes of the EU funded Horizon 2020 (H2020) research program particularly as applied to food security and sustainable agriculture and the one health approach, that links human to animal medicine.

One of the serious problems encountered in the early stages of farm animal proteomics was the lack of species genomes available for bioinformatics interpretation of proteomic data. This has been alleviated by the generation of genomes for more of the domestic animal species used for food production. However there has been a lack of investment in the key stage of annotating the genes in relevant genomes to proteins encountered in the animal. This is the next challenge to overcome to really establish the use of proteomics as an indispensable tool for research in farm animal and aquaculture science.

Our final meeting is in Milan and there can be no better location for disseminating the accumulated knowledge generated by the COST Action on Farm Animal Proteomics with the organization of the Farm Animal Proteomics 4th and Final Showcase meeting being seen in the framework of the events organized with the patronage of Padiglione Italia – EXPO 2015, (1st May – 31st October 2015), whose theme is 'feeding the planet. Energy for life'. In fact, scientific and technological aims of Farm Animal proteomics parallel those of world exposition, which will turn the world attention to the major social and technological improvement to increase safety, and quality, along the agro-food chain through innovative good practices. Italy is famous worldwide for its food from animal origin linked to tasteful and traditional cuisine as much as for its huge artistic and historical assets. Milan in particular is the heart of the Lombardy, which leads the Italian agricultural sector, the activities of which cover 69% of the national territory. Beside agro-food industry, Lombardy is also at the head of Italian biotechnological research. Seven Universities are also located here, with more than 150,000 students, together with several International research institutions, which makes Milan the leading hub for biomedical and biotechnological research in Italy, renowned for their innovations in the healthcare industry. Our meeting is hosted in the prestigious location of by Regione Lombardia, which is institutionally oriented toward promoting and supporting initiatives aimed to parallel innovation with food quality and safety, that is one of the central theme of our COST ACTION, where the results of this action can be revealed to a public that is very aware of these arguments.

We therefore welcome you all to Milan, Lombardia, Italy and to a bright future in farm animal proteomics research.

Part I
Oral communications

Perspective in comparative proteomics research

Andrea Urbani

Department of Experimental Medicine and Surgery, University of Rome 'Tor Vergata', Rome, Italy;
andrea.urbani@uniroma2.it
IRCCS-Fondazione S. Lucia, Rome, Italy
Centre of Investigation on Aging, Fondazione Università 'G.D'Annunzio', Chieti, Italy

Hippocrates, the Greek Father of Medicine, based his approach recognizing that human health, animal health and ecosystem health are part of a whole body. In fact, guiding the classical heroes (today scientists) we find the Chiron, the wisest of all centaurs is the mythological classical representation of an integrated view between man and the environment. Thus wisdom is achieved in the integration of man with the neighbouring nature. The main initiative within the framework of the World Proteomics is the definition of the Human Proteome, launched by the President of USA Mr. B. Obama in 2010. In the following years a number of associated actions have been initiated in different field of human health (1,2). However, so far only a small enthusiastic group of scientists have been pursuing the mythological Chiron in modern System Medicine (3,4). In fact the integrated body of evidence in proteomics investigations are providing the key molecular and analytical knowledge to achieve evidence based approach on these complex relationships. The analogical nature of many molecular objects, in particular proteins, in being part of a large molecular biosystem is a fundamental concept in proteomics investigations. These are open unsupervised studies which have been providing to the scientific community in life science an advanced tool to achieve a real novel knowledge of biological phenomena not necessarily link to an a priori hypothesis. These approaches have been bypassing canonical experimental designs which are often based on an arithmetic binary logic (5). However, the initial experiences in this filed have been employing the use of bioinformatics routine based on gene ontological associations. These networks of relationships are however built on knowledge-base databases which in fact are a reformatted representation of already published studies in the scientific literature mostly derived from evidences collected on human and murine samples. This represents a clear tautological limitation in the multifactorial interpretation of a posteriori investigations such as Proteomics data. In fact, the final results may not return the novel evidences which are the leitmotiv of Proteomics studies. In this scenario there is an urgent need to collecting a new body of evidences based on a comparative Proteomics vision. Such an intellectual process will provide a new insight, new ideas to be functionally explored within the fundamental observation of clinical phenotyping.

References

1   Hancock W, Omenn G, Legrain P, Paik YK. Proteomics, human proteome project,and chromosomes. J Proteome Res. 2011 Jan 7;10(1):210. http://dx.doi.org/10.1021/pr101099h. Epub 2010 Nov 29. PubMed PMID: 21114295.

2   Urbani A, De Canio M, Palmieri F, Sechi S, Bini L, Castagnola M, Fasano M, Modesti A, Roncada P, Timperio AM, Bonizzi L, Brunori M, Cutruzzolà F, De Pinto V, Di Ilio C, Federici G, Folli F, Foti S, Gelfi C, Lauro D, Lucacchini A, Magni F, Messana I, Pandolfi PP, Papa S, Pucci P, Sacchetta P; Italian Mt-Hpp Study Group-Italian Proteomics Association (www.itpa.it). The mitochondrial Italian Human Proteome Project initiative (mt-HPP). Mol Biosyst. 2013 Aug;9(8):1984-92. doi: 10.1039/c3mb70065h. Epub 2013 May 28. PubMed PMID: 23712443.

3   Roncada P, Modesti A, Timperio AM, Bini L, Castagnola M, Fasano M, Urbani A. One medicine--one health--one biology and many proteins: proteomics on the verge of the One Health approach. Mol Biosyst. 2014 Jun;10(6):1226-7. doi: 10.1039/c4mb90011a. Epub 2014 Apr 29. PubMed PMID: 24777557.

4   Bassols A, Turk R, Roncada P. A proteomics perspective: from animal welfare to food safety. Curr Protein Pept Sci. 2014 Mar;15(2):156-68. PubMed PMID: 24555902.

5   Urbani A, Castagnola M, Fasano M, Bini L, Modesti A, Timperio AM, Roncada P. Digital and analogical reality in proteomics investigation. Mol Biosyst. 2013 Jun;9(6):1062-3. doi: 10.1039/c3mb90013d. Epub 2013 Apr 30. PubMed PMID: 23629630.

Changes in the proteome of mastitis-causing *Escherichia coli* strains that affect pathogenesis

John Lippolis

Ruminant Diseases and Immunology Research Unit, National Animal Disease Center, Agricultural Research Service (ARS), United States Department of Agriculture (USDA), Ames, IA, USA; john.lippolis@ars.usda.gov

Escherichia coli is a leading cause of bacterial mastitis in dairy cattle. Milk is the environment in which bacteria must grow to establish an infection of the mammary gland. However, milk is not a rich growth media for bacteria. In fact, milk naturally contains many mechanisms to inhibit bacterial growth. How bacteria adapt to the mammary gland environment will likely be linked to the pathogenicity of the organism. We have used shotgun expression proteomics to determine the changes in protein expression when *E. coli* were grown in laboratory media compared to bacteria grown in whole fresh bovine milk. We found many proteins involved in the metabolism of lactose and various amino acids were up regulated when bacteria were grown in milk. We have also compared various strains of *E. coli* that are known to generate transient or persistent infection. Three persistent and three transient mastitis-derived strains of *E. coli* were compared using iTRAQ in a shotgun proteomics experiment. Expression data for 1127 proteins were determined. Of these, 27 proteins were associated with expression changes correlated with a difference in disease phenotype. Of particular interest were proteins that have been shown to be essential for bacterial swimming and swarming. Bacterial swimming and swarming assays showed that the strains from the persistent mastitis cases were significantly more mobile than the strains from the transient cases. This work identifies important protein expression differences between *E. coli* strains that cause a persistent versus a transient infection as well as demonstrates a corresponding difference in the associated bacterial motility phenotypes.

Colostrum protein uptake in neonatal lambs examined by descriptive LC-MS/MS

Lorenzo E. Hernández-Castellano[1,2], Anastasio Argüello[1], André M. Almeida[3,4], Noemí Castro[1] and Emøke Bendixen[5]*

[1]Department of Animal Science, Universidad de Las Palmas de Gran Canaria, Arucas, Spain

[2]Veterinary Physiology, Vetsuisse Faculty, University of Bern, Bern, Switzerland; lorenzo.hernandez@vetsuisse.unibe.ch

[3]Centro Interdisciplinar de Investigação em Sanidade Animal, FMV-UL; Instituto de Investigação Científica Tropical, Lisboa, Portugal

[4]Instituto de Biologia Experimental e Tecnológica and Instituto de Tecnologia Química e Biológica da UNL, Oeiras, Portugal

[5]Department of Molecular Biology and Genetics, Aarhus University, Aarhus, Denmark

Introduction

Colostrum is the first source of nutrition in neonatal ruminants, supplying not only nutrients, but having also a fundamental biological function, promoting immunoglobulin transfer from the dam to the newborn (Hernández-Castellano *et al.*, 2014a). Newborn ruminants are hypo gammaglobulinemic at birth, as the complexity of the ruminant placenta does not allow a sufficient transfer of immunoglobulins (Ig's) from the dam to the foetus (Lérias *et al.*, 2014). Therefore, newborn ruminants depend entirely on passive immunity transfer from the mother to the neonate, through the suckling of colostrum (Hernández-Castellano *et al.*, 2014b). The aim of this study is to describe the proteomes of sheep colostrum and lamb blood plasma after suckling, using SDS-PAGE for protein separation and in-gel digestion, followed by LC-MS/MS of resulting tryptic peptides for protein identification.

Material and methods

Sample collection and preparation

The study was based on individual analysis of plasma samples from 4 single partum lambs (Canarian dairy breed), studied as two experimental groups, as well as two samples from the standard pool of colostrum used for lamb feeding. The experiment took place at the experimental farm of the Veterinary Faculty of the Universidad de Las Palmas de Gran Canaria (Spain). During the experimental period (from birth up to 14 h after birth), the colostrum group (C group) received colostrum feeding at 2 h after birth. The non-colostrum group (NC group) was not fed at 2 h after birth. Blood samples were collected directly before feeding at 2 and 14 h after birth from the jugular vein in 2.5 ml tubes with K-EDTA. Blood was centrifuged at 2,190 g for 5 min at 4 °C and the obtained plasma was frozen at -80 °C until further analysis. Plasma and colostrum samples were analysed at Aarhus University (Aarhus, Denmark). A total of 200 µl from each sample were homogenized with 1 ml of TES buffer (10 mM Tris-HCl pH 7.6, 1 mM EDTA, 0.25 M sucrose) using an UltraTurrax® (T10 basic, IKA-Werke, Staufen, Germany) at 12,000 rpm. Homogenates were centrifuged at 10,000×g for 30 min at 4 °C to remove insoluble components. Protein concentration of the supernatant was determined with the Quick Start™ Bradford Protein Assay (Bio-Rad, Hercules, CA, USA), using BSA as standard reference (Bradford, 1976) and following manufacturer's instructions. Aliquots of 100 µg protein from each sample was recovered after precipitation with 6 volumes of ice-cold acetone (-20 °C) centrifuged at 15,000×g for 10 min at 4 °C. After that, 60 µg protein from the experimental groups (C and NC groups) at 14 h (2 biological replicates from each group) as well as the pooled colostrum used for the feeding experiment were prepared (2 technical replicates). Aliquots were resuspended in 20 µl TES buffer. The samples were boiled for 5 min in SDS sample buffer (7.5 ml) containing 500 mM DL-dithiothreitol (DTT), separated by SDS-PAGE using 10% (w/v) acrylamide gels and stained using Coomassie blue (RAPIDstain, G-Biosciences). Each lane was cut into 9 equal sized pieces, each of these were washed 3 times in milli-Q water and incubated two times for 15 min in 130 µl 50% acetonitrile, dehydrated in 130 µl acetonitrile for 15 min and equilibrated in 150 µl 0.1 M NH_4HCO_3 for 5 min before 150 µl acetonitrile was added. After 15 min, supernatants were removed and gel pieces were vacuum dried (SpeedVac®, Thermo Fisher Scientific, Waltham, MA, USA) for 20 min. In-gel digestions were performed by incubating the gel pieces with sequencing grade modified trypsin (Promega, Southampton, UK) in 50 mM NH_4HCO_3 at 37 °C for 16 h, using 400 ng trypsin for heavy loaded gel pieces, and 200 ng trypsin for more faint gel pieces. The resulting peptides were desalted using C18 StageTips (Thermo Scientific, Hvidovre, Denmark) and stored at -20 °C before LC–MS/MS analysis.

LC–MS/MS analysis and protein identification

LC–MS/MS analyses of gel pieces were performed on an EASY-nLC II system (Thermo Scientific) connected to a TripleTOF 5600 mass spectrometer (AB SCIEX) equipped with a NanoSpray III source (AB Sciex) and operated under Analyst TF 1.6 control. Tryptic peptides were dissolved in 15 µl of buffer A (0.1% formic acid), injected, trapped, and desalted isocratically on a ReproSil-Pur C18-AQ column (5 µm, 2 cm × 100 µm I.D;Thermo Scientific). Then, peptides were eluted from the trap column and separated on a ReproSil-Pur C18-AQ 3 µm capillary column (16 cm × 75 µm I.D) connected in-line to the mass spectrometer at 250 nl/min using a 50 min gradient from 5% to 35% of buffer B (0.1% formic acid and 90% acetonitrile), followed by a 10 min re-equilibration time in buffer A. The TripleTOF 5600 was run in positive ion mode using 2,500 V for ion spray, curtain gas at 30 psi, ion source gas at 5 and an interface heater temperature of 150 °C. The automated IDA method acquired up to 50 MS/MS spectra per cycle using 2.3 s cycle times and a mass exclusion window of 6 s. The peak lists were used to interrogate a combined bovine, ovine, and caprine database consisting of sequences from TrEMBL, Swiss-Prot and NCBInr (32,444 sequences), using Mascot 2.3.02 (Matrix Science). The search parameters were set to allow one missed trypsin cleavage site and propionamide as a fixed modification. The mass accuracy of the precursor and product ions were 15 ppm and 0.2 Da and the instrument settings were specified as ESI-QUAD-TOF. Only proteins with at least 2 unique peptides and a minimal protein score of 60 were considered for identification (Boersema et al., 2008).

Results and discussion

In order to investigate which proteins can be detected in colostrum and plasma using shotgun proteomics, we used an approach based on SDS-PAGE separation and LC-MS/MS based protein identification. Only proteins detected in both biological replicates from each sample type (C and NC groups at 14 h after birth as well as the colostrum pool used) were considered significant and unique observations. Analyses of the pooled colostrum which was used in our feeding experiment, allowed for the identification of 70 proteins, while 64 and 97 proteins could be observed in plasma from C and NC groups, respectively. The proteins observed in plasma from colostrum animals represent both original plasma proteins as well as the subset of proteins that are transferred from colostrum to plasma. Figure 1 presents an overview of the distribution of proteins across colostrum and plasma. It is important to notice that 29 observed proteins were detected in all 3 studied groups (colostrum group and C and NC lambs). Moreover, comparing these proteome profiles clearly demonstrated that a wide range of immunoglobulins originating from colostrum are transferred to plasma, while a large majority of colostrum specific proteins were not detectable in plasma 12 h after suckling, indicating that transport of proteins from colostrum, over the gut epithelial layer and into plasma is a controlled and selective process, although mechanisms responsible for selective transfer are not yet fully understood (Johnson *et al.*, 2007). Another interesting observation provided by our direct comparison of body fluid proteomes is that a much larger range of well-known plasma proteins and also intracellular proteins (i.e. fructose –biphosphate aldolase and carbonic anhydrase) are detectable in plasma from the animals that lacked access to colostrum when compared to the plasma of colostrum fed lambs. This is likely due to the fact that peptide selection in shotgun-based LC-MS/MS analyses is greatly biased by the relative abundance of specific proteins and peptides in complex biological samples. Thus, in plasma collected after colostrum suckling, the high concentration of immunoglobulins present in plasma clearly outnumber and overshadow detection of the lower abundance plasma-protein in the background as previously suggested (Petersen *et al.*, 2013).

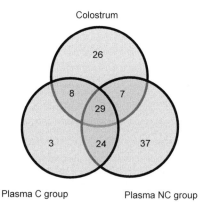

Figure 1. Venn diagram of colostrum pool as well as plasma from colostrum lambs (C group) and non-colostrum (NC group) lambs at 14 h after birth.

Conclusions

The results of this study described the presence of 70 proteins in the ovine colostrum proteome. Moreover, 64 and 97 proteins were detected in the C and NC groups, respectively. Understanding the uptake and effect of colostrum is important for future reduction of lamb mortality rates related to artificial rearing, therefore with possible benefits for sheep farmers. In this study 29 proteins were detected in colostrum and lamb plasma from both groups. Further proteomic studies will be necessary, particularly using the Selected Reaction Monitoring (SRM) approach, in order to increase the general knowledge about the role of colostrum in the passive immune transfer.

Acknowledgements

Authors acknowledge financial support from the European Cooperation in Science and Technology (COST), through the FA-1002 COST action, the Formación del Profesorado Universitario (FPU) program (Ministry of Education, Madrid, Spain) as well as the program Ciência 2007 from *Fundação para a Ciência e a Tecnologia* (Lisbon, Portugal).

References

Boersema, P.J., Aye, T.T., van Veen, T.A., Heck, A.J. and Mohammed, S., 2008. Triplex protein quantification based on stable isotope labeling by peptide dimethylation applied to cell and tissue lysates. Proteomics 8: 4624-4632.

Bradford, M.M., 1976. A rapid and sensitive method for the quantitation of microgram quantities of protein utilizing the principle of protein-dye binding. Analytical Biochemistry 72: 248-254.

Hernández-Castellano, L.E., Almeida, A.M., Castro, N. and Argüello, A., 2014a. The colostrum proteome, ruminant nutrition and immunity: a review. Current protein and peptide science 15: 64-74.

Hernández-Castellano, L., Almeida, A., Ventosa, M., Coelho, A., Castro, N. and Arguello, A., 2014b. The effect of colostrum intake on blood plasma proteome profile in newborn lambs: low abundance proteins. Bmc Veterinary Research 10: 85.

Johnson, J.L., Godden, S.M., Molitor, T., Ames, T. and Hagman, D., 2007. Effects of feeding heat-treated colostrum on passive transfer of immune and nutritional parameters in neonatal dairy calves. Journal of Dairy Science 90: 5189-5198.

Lérias, J.R., Hernández-Castellano, L.E., Suárez-Trujillo, A., Castro, N., Pourlis, A. and Almeida, A.M., 2014. The mammary gland in small ruminants: major morphological and functional events underlying milk production – a review. Journal of Dairy Research 81: 304-318.

Petersen, L.J., Sorensen, M.A., Codrea, M.C., Zacho, H.D. and Bendixen, E., 2013. Large pore dermal microdialysis and liquid chromatography-tandem mass spectroscopy shotgun proteomic analysis: a feasibility study. Skin Research and Technology 19: 424-431.

Proteomic investigation in milk whey from healthy and mastitis affected Jafarabadi buffaloes (*Bubalus bubalis*)

André M. Santana[1]*, Daniela G. Silva[2], Funmilola Thomas[1], Richard J.S. Burchmore[3], José J. Fagliari[2] and Peter D. Eckersall[1]

[1]*Institute of Biodiversity, Animal Health and Comparative Medicine, University of Glasgow, Bearsden Rd, G61 1QH, Glasgow, United Kingdom; andre.santana@glasgow.ac.uk*

[2]*Department of Veterinary Clinics and Surgery, FCAV, UNESP, Sao Paulo, Brazil*

[3]*Glasgow Polyomics Facility, University of Glasgow, G12 8QQ, Glasgow, United Kingdom*

Introduction

Mastitis is probably the most costly of the infectious diseases in farm animals and is becoming an increasing problem for the dairy industry in many countries, causing major economic losses, especially when present in a subclinical form when clear clinical signs such as swelling, redness, tenderness of the udder and clots in the milk are absent. Although buffaloes are considered less susceptible than cattle for clinical mastitis, studies have shown similar subclinical mastitis occurrence in these species (Bastos and Birgel, 2011). In this sense, reliable detection of mastitis is necessary for controlling the disease and for monitoring milk quality. Tests for indicators of inflammation can be used to screen quarters for intramammary inflammation and be useful as diagnostic and prognostic parameters, to indicate the intensity of inflammation and to monitor recovery from the disease. In this context, acute phase proteins (APP's) have shown positive correlation with the severity of mastitis infection in cows (Eckersall *et al.*, 2001; Pyörälä *et al.*, 2011) and can be used as a powerful diagnostic tool. Therefore, this study investigates, using 2-DE approach, modifications of the milk whey proteome profile in Jafarabadi buffaloes with mastitis, in order to identify possible additional biomarker for this disease.

Material and methods

Milk samples were collected from 82 mammary quarters of healthy Jafarabadi buffaloes (negative bacteriology, SCC<100.000 cels/ml, negative CMT) and from 97 mammary quarters from Jafarabadi buffaloes with subclinical mastitis (positive bacteriology, SCC>100.000 cels/ml, positive CMT) in a farm located in Sao Paulo State, Brazil. Before collecting the samples, physical examination of the mammary gland, as well as California Mastitis Test (CMT) was conducted. After disinfection of teats, milk samples from each mammary quarter were collected for microbiological isolation and for somatic cell count (SCC). Milk whey samples were obtained by addition of 5% renin solution followed by centrifugation, so that solids and caseins were removed before performing two-dimensional electrophoresis (2-DE). Milk whey total protein concentration were measured by Bradford assay. To define which samples were selected for 2-DE analyses, the milk whey haptoglobin concentration was measured. Three whey samples with low (healthy buffaloes) and high (mastitic buffaloes) haptoglobin concentrations where selected for separation of proteins by 2-DE which was accomplished using 11 cm, pH 3-10 IPG Strips (Bio-Rad Labs, Hemel Hempstead, UK) followed by SDS-PAGE on 4-15% polyacrylamide gels and stained with Coomassie blue. Protein spots will be excised and subjected to tryptic in-gel digestion and analysed by LC-MS/MS.

Results

Total protein concentrations ranged from 1.11 to 1.30 mg/ml and 1.74 to 2.50 mg/ml in healthy and mastitic buffaloes, respectively. Haptoglobin concentrations ranged from 0.10 to 0.37 mg/l and 5.24 to 10.04 mg/l in healthy and mastitic buffaloes, respectively. For a precise comparison between healthy and mastitic buffaloes, equal protein loading of 200 µg were used, in all samples, for 2-DE protein separation. The 2-DE gels healthy and mastitic buffalo milk whey samples (Figure 1) showed similar protein composition though some differences in patterns and expression levels

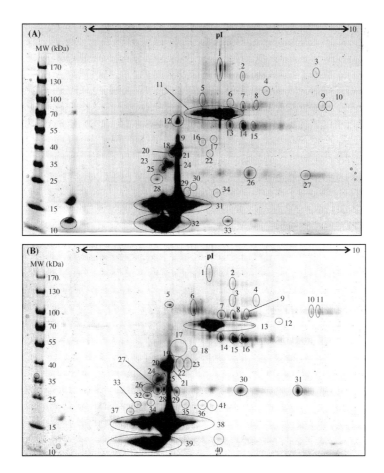

Figure 1. 2-DE of milk whey sample of healthy (A) and mastitic (B) buffalo with identification of protein spots that will be excised for proteomic analysis by LC-MS/MS.

were observed. For instance, 7 selected spots from healthy animal samples (3, 12, 17, 18, 22, 29 and 30 – Figure 1A) were not found in mastitic animals. On the other hand, 13 selected spots from mastitic animals (3, 5, 17, 22, 23, 26, 28, 29 and 33 to 37 – Figure 1B) were not found in healthy animals. Based on previous reports in buffaloes (D´Ambrosio *et al.*, 2008) and bovines (Alonso-Fauste *et al.*, 2012), protein spots from healthy (1, 5 to 10, 13 to 15, 23 and 24 – Figure 1A) and mastitic animals (1, 5-11, 14-16, 24 and 25 – Figure 1B) are likely to be host-defence proteins such as IgG's, IgM, lactoperoxidase, lactoferrin and α_1-acid glycoprotein. Other spots from healthy (25 to 27 – Figure 1A) and mastitic animals (26 to 31 – Figure 1B) could be caseins remaining after rennin treatment. Spots 11, 31 and 32 (healthy animals – Figure 1A) and 13, 38 and 39 (mastitic animals – Figure 1B) are likely to be the high abundance proteins albumin, β-lactoglobulin and α-lactalbumin, respectively.

Conclusion

2-DE separated buffalo milk whey into its constituent proteins, allowing a comparison of protein spots between healthy and mastitic buffaloes. The protein profile showed differences in patterns and expression levels, which will enable future identification of proteins by LC –MS/MS to provide valuable information in identification of possible biomarkers for mastitis in buffaloes.

Acknowledgements

FAPESP (Process number: 2013/26498-5) and CNPq are thanked for financial support.

References

Alonso-Fauste, I.; Andrés, M.; Iturralde, M.; Lampreave, F.; Gallart, J.; Álava, M. A., 2012. Proteomic characterization by 2-DE in bovine serum and whey from healthy and mastitis affected farm animals Journal of proteomics 75, 3015-3030.

Bastos, P. A. S. and Birgel, e. h., 2011. Milk of Murrah buffaloes bred in São Paulo (Brazil): influence of age, lactation phase, time of milking and bacterial isolation in the physical chemical and cell composition. Journal of Continuing Education in Animal Science 9, 6-13.

D´Ambrosio, C.; Arena, S.; Salzano, A. M.; Renzone, G.; Ledda, L.; Scaloni, A., 2008. A proteomic characterization of water buffalo milk fractions describing PTM of major species and the identification of minor componentes involved in nutriente delivery and defense against pathogens. Proteomics 8, 3657-3666.

Eckersall, P. D.; Young, F. J.; Mccomb, C.; Hogarth, C. J.; Safi, S.; Weber, A.; Mcdonald, T.; Nolan, A. M.; Fitzpatrick, J. L., 2001. Acute phase proteins in serum and milk from dairy cows with clinical mastitis. Veterinary Record 148, 35-41.

Pyörälä, S.; Hovinen, M.; Simojoki, H.; Fitzpatrick, J.; Eckersall, P. D.; Orro, T., 2011. Acute phase proteins in milk in naturally acquired bovine mastitis caused by different pathogens. Veterinary Record 168, 535.

Bovine LDL and HDL proteome in health and inflammatory state

Milica Kovačević Filipović[1]*, Emily L. O'Reilly[2] and P. David Eckersall[2]

[1]Faculty of Veterinary Medicine, University of Belgrade, Serbia; milkovac@yahoo.com
[2]Institute of Biodiversity, Animal Health & Comparative Medicine, University of Glasgow, United Kingdom

Introduction

Lipoproteins are essential for lipid metabolism. In bovine plasma, the most abundant are high density lipoproteins (HDL) which have a role in reverse cholesterol transport and are a major source of essential fatty acids. Low density lipoproteins (LDL) carry cholesterol to extra-hepatic tissues. Apo A-I, A-II, E are major apo-lipoproteins associated with HDL and Apo B100 and E are associated with LDL. Recent proteomic analysis in humans and bovines demonstrate that as well as these well-known apo-lipoproteins, multiple other proteins could be associated with HDL (Gordon *et al.*, 2010, Della Donna *et al.*, 2012) and LDL (Davidsson *et al.*, 2005). Some of these proteins have roles in immunodulation (Gordon *et al.*, 2010). Inflammation (per se) has a qualitative and quantitative influence on lipid metabolism and influences HDL structure and function (Gordon *et al.*, 2010). Fewer investigations have been performed on changes in LDL metabolism, structure and function during inflammation.

The aim of this study was to compare HDL and LDL proteomes in healthy cows and cows with an acute phase response, in order to give novel information regarding the relationship between inflammation and metabolism.

Material and methods

Pooled sera of healthy cows with undetectable haptoglobin (Hp) and acute phase (AP) cows with elevated Hp=0.87 g/l, (n=6), were subjected to selective pre-fractionation for lipoprotein, by gel filtration following Gordon *et al.* (2010). Sera (400 µl each) were applied to a Superdex 200 gel filtration column (GE Healthcare, UK). Samples were processed at a flow rate of 0.3 ml/min in standard Tris buffer using an Akta chromatography system (GE Healthcare, UK). Eluate was collected in 20 fractions (600 µl each). Protein and cholesterol concentration after gel chromatography were determined using standard laboratory protocols.

Lipid precipitation was performed in each fraction with calcium silicate hydrate – CSH (Sigma, UK) at 25 °C, according to Gordon *et al.* (2010). Pellets containing apo-lipoproteins were obtained after 30 minutes of vortexing, 2 minutes of centrifugation and three washings in AB. Precipitated apo-lipoproteins were extracted from CSH pellets using Laemmli buffer (100 µl) and used for SDS PAGE electrophoresis. To verify the success of CSH treatment, cholesterol levels were measured in supernatants.

Protein identification by mass spectrometry (MS) and peptide mass fingerprinting of protein bands from the SDS-PAGE gels, after trypsin digest was performed in the Glasgow Polyomics Facility, according to the procedure described in Braceland et al. (2013). MS analysis were done only with proteins extracted from CSH precipitate. One LDL and one HDL fraction were analyzed. Also, bands with medium and low molecular weight proteins from LDL and HDL fraction obtained after electrophoresis were analyzed.

Results

After gel chromatography, LDL and HDL fractions were defined according to cholesterol content. Analysis of cholesterol in supernatants after CSH precipitation revealed that approximately 80 to 90% of this molecule precipitated (data not shown). One dimension SDS PAGE separation of proteins from LDL and HDL fractions before and after CSH treatment have shown that protein bands did not differ substantially indicating that, although proteins of interest were concentrated, probably contamination with proteins not belonging only to LDL and HDL was present and data obtained after MS analysis should be interpreted with precautions. The diversity of proteins found in chosen LDL fractions and bands could be seen on Table 1.

MS analysis revealed that HDL in both, healthy and acute phase cows contain apolipoprotein AI in all analyzed fractions (Table 2). As in the case of LDL, analysis of whole HDL pellet is less powerful in detection of different proteins than analysis of separated bands (Table 2).

Discussion

Proteomic analysis of apo-lipoprotein in serum is hindered by the presence of high abundance protein such as albumin and γ-globulins. This can be addressed by selective pre-fractionation for lipoprotein by gel filtration and affinity absorption on CSH (Gordon et al., 2010). In this project, these methods were applied to bovine serum for the first time. MS analysis revealed that putative LDL fractions contained α2 macroglobulin and ceruloplasmin in both groups and Hp in acute phase cows. HDL fraction contained ceruloplasmin in healthy and SAA in both groups. It is known that SAA in plasma is bound to HDL and even small concentrations of these protein in healthy cow sera were detected after low molecular weight bands analysis. According to these results it was concluded that LDL and HDL proteins were adequately separated using gel chromatography. The unusual finding is that among known apo-lipoproteins, only Apo A-I was detected in both, LDL and HDL fractions. Apo A-I is typical HDL component, although recent literature described Apo A-I as a minor component of LDL also (Von Zychlinsky et al., 2014). Other proteins identified in LDL and HDL fractions were different complement components, proteins involved in hemostasis (trombospondin, kalikrein sensitive protein), proteinase inhibitors (α1 anti-plasmin, α trypsin inhibitor, α1 anti-chymotrypsin), extracellular matrix proteins (vitronectin, tenascins) and transport proteins (transthyrethrin). Among these proteins, no clear connection with beneficial or detrimental effect of LDL and HDL change during inflammatory state could be found. The only obvious, consistent difference between healthy and acute phase cows was that the later had more proteins

Table 1. Proteins identified by MS in LDL fractions obtained after gel chromatography.

Partial bovine LDL proteom			
Healthy	**Acute phase**	**Healthy**	**Acute phase**
fraction 8	fraction 36	fraction 8, 70 kD Band	fraction 36, 70 kD Band
Apolipoprotein A-I	Apolipoprotein A-I	Apolipoprotein A-I	/
α2 macroglobulin	α2 macroglobulin	α2 macroglobulin	α2 macroglobulin
Ceruloplasmin	Ceruloplasmin	Complement comp 9	Haptoglobin
α1 anti-proteinase	Haptoglobin	α1 anti-proteinase	α1 anti-chymotrypsin
	a-fetoprotein	α trypsin inhibitor	Serpin A3
	Kalikrein sensitive protein		Vitamin D banding protein
	ATP-binding casette		Pancreatic elastase inhibitor
	Trombospondin		
		fraction 8, 25 kD Band	fraction 36, 25 kD Band
		Apolipoprotein A-I	Apolipoprotein A-I
		α2 macroglobulin	α2 macroglobulin
		Complement comp 4	Complement comp 3
		PPRγ	Malate dehidrogenase
		Multidrug resistance associated protein 4	Kinectin
		fraction 8, 10 kD Band	fraction 36, 10 kD Band
		Apolipoprotein A-I	Haptoglobin
		α2 macroglobulin	Tenascin
		Complement C4 BP	Transthyrethrin
		PPRγ	
		Multidrug resistance associated protein 4	

connected with both LDL and HDL fractions, and further analysis of importance of this findings should be performed. Analysis of gel chromatography fractions on SDS PAGE electrophoresis before and after CSH precipitation of LDL and HDL demonstrated that CSH protocol was not optimized because practically all serum proteins precipitated. Optimization of this step is critical for further analysis of the LDL and HDL proteome.

From data that we obtained, we can conclude that the method used, at least partially, discriminated different lipoprotein fractions and concentrated different proteins associated with LDL and HDL in both, healthy and diseases cows. Also, acute phase cows have abundant proteins connected to

Table 2. Proteins identified by MS in HDL fractions obtained after gel chromatography.

Partial bovine HDL proteom			
Healthy	**Acute phase**	**Healthy**	**Acute phase**
fraction 11	fraction 39	fraction 11, 10 kD band	fraction 39, 10 kD band
Apolipoprotein A-I	Apolipoprotein A-I	Apolipoprotein A-I	Apolipoprotein A-I
Ceruloplasmin	Ceruloplasmin	Complement comp 3,4	Complement comp 3,4
Complement comp 3,4,5	Complement comp 3,4,5	Serum amyloid A	Serum amyloid A
α-fetoprotein	α-fetoprotein	Serum amyloid P	Serum amyloid P
α1 anti-plasmin	Serum amyloid A	Anti-testosteron antibody	Anti-testosteron antibody
		Olfactomedin	Olfactomedin
		MMP-13	MMP-13
			Tenascin
			Transthyrethrin

both, LDL and HDL fractions that, according to their known function, probably influence lipid metabolism, or have some immuno-modulatory properties.

Acknowledgement

This works had financial support from short term scientific mission realized in frame of COST Action FA 1002 (COST-STSM-FA1002-290913-035336).

References

Braceland, M, Bickerdike, R, Tinsley, J Cockerill D, Eckersall PD, 2013. The serum proteome of Atlantic salmon, Salmo salar, during pancreas disease (PD) following infection with salmonid alphavirus subtype 3 (SAV3). Journal of Proteomics 6, 423-36.

Davidsson P, Hulthe J, Fagerberg B, Olsson BM, Hallberg C, Dahllöf B, Camejo G, 2005. A proteomic study of the apolipoproteins in LDL subclasses in patients with the metabolic syndrome and type 2 diabetes. Journal of Lipid Research. 46, 1999-2006.

Della Donna L, Bassilian S, Souda P, Nebbia C, Whitelegge JP, Puppione DL, 2012. Mass spectrometric measurements of the apolipoproteins of bovine (Bos taurus) HDL. Comparative Biochemistry Physiology Part D Genomics Proteomics 7, 9-13.

Gordon SM, Deng J, Lu LJ, Davidson WS, 2010. Proteomic characterization of human plasma high density lipoprotein fractionated by gel filtration chromatography. Journal of Proteome Research 1, 5239-49.

Von Zychlinski A, Williams M, McCormick S, Kleffmann T. 2014. Absolute quantification of apolipoproteins and associated proteins on human plasma lipoproteins. Journal of Proteomics 106, 181-90.

Application of top-down proteomics in detecting biomarkers in porcine saliva samples

María Fuentes-Rubio[1]*, Monica Sanna[2], Tiziana Cabras[2], Massimo Castagnola[3], Federica Iavarone[3], F. Tecles[1], J. Ceron[1] and Irene Messana[2]

[1]*Department of Animal Medicine and Surgery, University of Murcia, 30100, Murcia, Spain; mfrvet.84@gmail.com*
[2]*Department of Life and Environmental Sciences, University of Cagliari, Italy*
[3]*Department of Biochemistry and Clinical Biochemistry, Catholic University of Rome, Italy*

Objectives

In recent years, the use of proteomics for the study of biomarkers has been one of the most used techniques (Gao *et al.*, 2005) and it has provided promising results in the characterization of salivary proteins and peptides in pigs. The two-dimensional gel electrophoresis (2DE) map of porcine saliva (Gutiérrez *et al.*, 2011) has been used to identify biomarkers of disease in growing pigs under field conditions (Gutiérrez *et al.*, 2013) and under stress conditions (Fuentes-Rubio *et al.*, 2013).

However, different proteomics strategies based on top-down platforms can allow to evaluate different aspects of the variation of porcine salivary proteome under stress conditions (Messana *et al.*, 2013).

Material and methods

Saliva samples were taken from a group of 20 healthy crossbreed growing pigs by introducing a small sponge in the pigs mouth with the help of a metal rod as previously described (Fuentes *et al.*, 2011). Samples were placed in collection devices (Salivette, Sarstedt AG & Co., Nümbrecht, Germany) and kept ice until the arrival at the laboratory. The samples were centrifuged at 4,000×*g* for 8 min and then were kept at -80 °C until analysis. All animals were subject to a stress induction consistent in the immobilization of the pigs during 1 min with a nasal snare following the procedure described in the literature (Fuentes *et al.* 2011) as representative of an acute stressor for pigs (Geverink *et al.*, 2001; Rushen *et al.*, 1991). Four salivary samples were taken from each animal: before the stress induction as a control sample (TB); just after the stress induction (T0); 15 min (T15) and 30 min after the stress induction. The research protocol was approved according to the European Council Directives regarding the protection of animals used for experimental purposes.

Two aliquots were made from every sample. One aliquot was used for biochemical analysis and the other was mixed 1:1 v/v with 0.2% trifluoroacetic acid (TFA) on ice, and centrifuged 5 min at 8,000×*g* at 4 °C. The precipitated was discarded and 400 µl of the total volume was frozen at -80 °C.

Each aliquot was thawed and centrifuged at 13,000 rpm during 5 min at 4 °C. 100 µl of the soluble fraction was submitted to analysis. Low-resolution HPLC-ElectroSpray Ionization-Ion Trap-Mass Spectrometry (HPLC-ESI-IT-MS) measurement was performed by a Surveyor HPLC system (ThermoFisher, San Jose, CA, USA) connected by a T splitter to a photo diode-array detector and to a LCQ Advantage Mass Spectrometer. The RP-HPLC-ESI-IT-MS analysis followed a modified protocol previously published (Cabras *et al.*, 2009, 2013). After the analysis, a total ion current (TIC) chromatogram (Figure 1) of whole saliva was obtained for each sample. The software Xcalibur 2.07 (ThermoFisher, San Jose, CA, USA) was used to analyse the files. Deconvolution of averaged ESI mass spectra was automatically performed by MagTran 1.0 software (Zhang and Marshall, 1998) and allowed to obtain the experimental mass values of peptides and proteins present in the sample.

Peptides already characterized in porcine saliva (Fanali *et al.*, 2008) were searched in the profile by using the XIC (eXtracted Ion Current) procedure, and the area of the XIC peaks measured for quantification. The experimental average mass value of the mono charged ion ([M+H]$^+$), the elution time and the peak area of each peptide were registered in spread sheet (Microsoft Excel, Microsoft Corp., Redmond, WA, USA).

For structural characterization 20 µl of eleven selected samples were submitted to high-resolution HPLC-ESI-MS/MS analysis by an LTQ Orbitrap XL apparatus (ThermoFisher, San Jose, CA, USA),

The software Proteome Discoverer 1.4 (Thermo, San Jose, CA, USA) was applied to analyse the MS/MS spectra and determine the sequences of peptides with masses lower than 10,000 Da. The following search parameters were applied: no missed cleavage sites were allowed and a fragment mass tolerance of 0.6 Da was permitted. The allowed Post-Translational Modifications (PTMs) were N-Terminal Modification and a variable modification of phosphorylation (S-Phospho). The

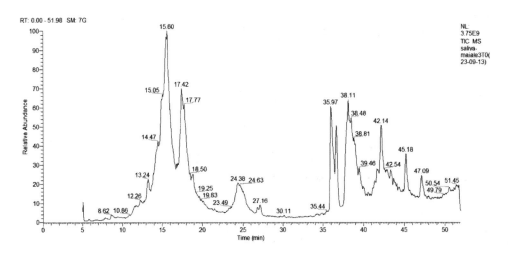

Figure 1. Typical total ion current (TIC) chromatogram of porcine whole saliva sample.

experimental MS/MS spectra were compared to the theoretical ones generated by using the web tool MS Product (Protein Prospector, http://prospector.ucfs.edu) in order to confirm the sequences suggested by Proteome Discoverer.

The software GraphPad Prism (version 4.0, GraphPad Software Inc., La Jolla, CA, USA) was used for statistical analysis. ANOVA Statistical analysis was considered significant when $P<0.05$.

Results and conclusion

The Proteome Discoverer results confirmed the presence of known peptides generated by the three isoforms of a PRP pro-protein (Swiss-Prot data bank: Q95JC9-1, Q95JC9-2 and Q95JC9-3) (Fanali *et al.*, 2008), and also evidenced that the majority of the experimental masses were compatible with multiple fragments generated by the Proline-Rich Protein (PRP) precursors. Even though the TIC chromatograms of the samples showed similar profiles from a qualitative point of view, statistical analysis evidenced some differences of the XIC peak areas of several protein fragments before and after stress induction, even though not statistically significant (Table 1).

The high fragmentation detected could be rationalised according to different possibilities: (1) the stress can origin a major release of the isomers of the PRP protein Q95JC9, so it is possible to found more fragments of these; (2) the stress can origin a major activity of proteases in saliva so,

Table 1. Protein fragments identified before (TB) and after stress (T0, T15, T30).[1]

Name	Mr Theor (Da)	Mr Exp (Da)	Area of the XIC peak in arbitrary units (mean value±SEM)			
			TB	T0	T15	T30
PRP-SP-A	6,156.5	6,156.1	6.67E+06±4.11E+06	2.10E+08±1.59E+08	NF	NF
PRP-SP-B	1,904.2	1,903.8	1.49E+09±1.02E+09	6.12E+09±2.16E+09	4.26E+08±2.25E+08	5.82E+08±2.52E+08
PRP-SP-C	881.0	NF	NF	NF	NF	NF
PRP-SP-D	1,510.7	1,509.9	5.16E+07±3.54E+07	3.62E+08±2.18E+08	1.36E+08±1.32E+08	2.05E+07±2.05E+07
PRP-SP-E	2,733.3	2,732.7	3.11E+08±2.31E+08	1.80E+09±1.05E+09	1.58E+08±1.16E+08	2.33E+07±1.21E+07
PRP-SP-F	1,920.2	1,920.2	6.87E+07±6.66E+07	3.40E+08±1.77E+08	2.56E+07±1.71E+07	1.14E+07±6.20E+06
PRP-SP-G	2,166.4	2,166.7	1.49E+08±9.77E+07	5.46E+08±2.12E+08	4.66E+07±2.37E+07	3.49E+07±1.97E+07
PRP-SP-H	2,012.3	2,012.8	1.76E+08±1.25E+08	7.29E+08±3.08E+08	4.24E+07±2.95E+07	4.70E+07±2.48E+07
PRP-SP-I	3,790.0	3,789.2	4.02E+07±2.39E+07	2.45E+08±1.15E+08	7.95E+06±4.41E+06	1.83E+07±9.37E+06
PRP-SP-L	1,859.1	NF	NF	NF	NF	NF
PRP-SP-M	1,652.9	1,652.4	1.23E+08±8.20E+07	5.16E+08±2.07E+08	4.07E+07±2.72E+07	3.34E+07±1.63E+07
PRP-SP-N	1,223.4	1,222.8	4.43E+07±3.12E+07	1.25E+08±6.17E+07	2.22E+07±1.28E+07	6.18E+06±3.50E+06

[1] M_r Theor: molecular weight theoretical; M_r Exp: molecular weight experimental; NF: not found.

the proteins were more fragmented; (3) a different proteolysis occurred between the collection and stabilization of the samples. A more detailed analysis of the results it is necessary to understand their biological implications because the behaviour of protein fragments was not homogeneous.

References

Cabras, T. *et al.*, 2009. Age-Dependent Modifications of the Human Salivary Secretory Protein Complex. Journal of Proteome Research 8, 4126-4134.

Cabras, T. *et al.*, 2013. Significant Modifications of the Salivary Proteome Potentially Associated with Complications of Down Syndrome Revealed by Top-down Proteomics. Mol Cell Proteomics 12, 1844-1852.

Fanali, C., *et al.*, 2008. Mass spectrometry strategies applied to the characterization of proline-rich peptides from secretory parotid granules of pig (*Sus scrofa*). J. Sep. Sci. 31, 516-522.

Fuentes, M. *et al.*, 2011. Validation of an automated method for salivary alpha-amylase measurements in pigs and its application as a biomarker of stress. J Vet Diagn Invest 23, 282-287.

Fuentes-Rubio, M. *et al.*, 2014. Porcine salivary analysis by two-dimensional gel electrophoresis in three acute stress models: a pilot study. Can J Vet Res 78, 127-132.

Gao, J. *et al.*, 2005. Biomarker discovery in biological fluids. Methods 35, 291-302.

Geverink, N.A. *et al.*, 2002. Individual differences in behavioral and physiological responses to restraint stress in pigs. Physiol Behav 77, 451-457.

Gutiérrez, A.M. *et al.*, 2011. Proteomic analysis of porcine saliva. Vet J 187, 356-362.

Gutiérrez, A.M. *et al.*, 2013. Detection of potential markers for systemic disease in saliva of pigs by proteomics: A pilot study. Vet Immunol Immunopathol 151, 73-82.

Messana, I. *et al.*, 2013. Unraveling the different proteomic platforms.J Sep Sci. 36, 128-139.

Patamia, M., *et al.*, 2005. Two proline-rich peptides from pig (*Sus scrofa*) salivary glands generated by pre-secretory pathway underlying the action of a proteinase cleaving Pro Ala bonds. Peptides 26, 1550-1559.

Rushen, J. and Ladewig, J., 1991. Stress-induced hypoalgesia and opioid inhibition of pigs' responses to restraint. Physiol Behav 50, 1093-1096.

Zhang, Z. and Marshall, A.G., 1998. A universal algorithm for fast and automated charge state deconvolution of electrospray mass-to-charge ratio spectra. J Am Soc Mass Spectrom 9, 225-233.

Proteomic characterization of molecular changes in saliva proteins in stressed pigs

Lourdes Criado[1]*, Laura Arroyo[1], Daniel Valent[1], Anna Bassols[1] and Ingrid Miller[2]
[1]Dept. Bioquímica,Universitat Autònoma de Barcelona, Spain, lourdescm90lc@gmail.com
[2]Dept. of Biomedical Sciences, University of Veterinary Medicine Vienna, Austria

Introduction

Saliva contains a number of proteins that may be useful as biomarkers of health and disease and can be easily obtained from large numbers of animals in a non-invasive, stress-free way (Gutiérrez *et al.*, 2011).

Saliva has been used to measure classical acute stress markers such as cortisol, alpha-amylase or chromogranin A (Obayashi, 2013). These parameters are sensible indicators of acute stress, but have several disadvantages. For these reasons we decided to study the protein composition in saliva in two groups of pigs subjected to short and long transport, as conditions of low and high-stress level. Furthermore, information about serum and saliva biochemistry parameters as well as post-mortem meat quality was available from this experiment, giving us the opportunity to undertake the search for new stress markers using a gel-based proteomic approach.

Objectives

The aim of this study was to investigate the changes in the protein composition of porcine saliva from pigs after short/easy or long/winding road transport from the farm to the slaughter, which represented conditions of no-stress and high degree of stress, respectively.

The goals of this study were: a) to analyze the changes in protein composition by means of one-dimensional SDS-PAGE (1-DE) and two-dimensional electrophoresis (2-DE); and b) to study the efficiency of a new fluorescent probe called T-Red310 for protein labelling.

This project was carried out in the laboratory of Ingrid Miller at the Veterinärmedizinische Universität (Vienna) thanks to a COST-Short Term Scientific Mission (FA 1002).

Materials and methods

Transport of pigs and sample collection

Pigs were divided in two groups: the control, non-stressed pigs (n=6) were transported to the slaughterhouse in a truck for a short time by an easy road (5 minutes). The second, stressed group (n=6), were transported to the slaughterhouse to the same slaughterhouse after a long transport (2 hours) by a winding road in the countryside.

Samples were collected before transport (the evening before) when the animals were fed *ad libitum*, and after transport where all of them had been fasted over-night (before and during transport), using a Salivette™ device.

T-Red labelling:

- Total protein content was determined using Bradford. For 1-DE 2,5 µg of protein was diluted to 10 µl with appropriate coupling buffer and labelling was performed by adding 1 µl of the T-Red310, followed by incubation on ice and in the dark for 30 minutes, and stopping the reaction with 1 µl of Lysine under similar conditions for 10 minutes.
- For 2-DE 25 µg of protein was diluted to 10 µl with appropriate buffer and were labelled with T-Red310 as described above and 28 µl of buffer were added.

1-DE:

- Samples were reduced with 12 µl of sample buffer containing DTT at 95 °C during 5 minutes, centrifuged for 5 minutes at 14,000 rpm and 12 µl of glycerol 70% was added. Proteins were separated on 140×140×1.5 mm gradient gels (10-15%) in a *Hoefer* SE600 vertical electrophoresis chamber and protein bands detected on a fluorescence scanner (Typhoon 9400).

2-DE:

- Selected samples were analyzed by 2-DE. First dimension was run under reducing and denaturing conditions in IPG homemade strips in a Multiphor II electrophoresis chamber (GE Healthcare) in non-linear gradient (pH 4-10). Strips were treated with DTT and iodoacetamide and the second dimension was performed as described for 1-DE. Protein spots were detected by a fluorescence scanner (Typhoon 9400).
- The same gels were also stained with MS-compatible silver stain to compare with the fluorescent label.
- In order to compare stressed with non-stressed pigs, densitometry of some 1-DE bands was performed with the ImageQuant T.L. software (GE Healthcare).

Results and discussion

1-DE protein profiles from some of the animals belonging to the long transport group, before and after transport, are shown in Figure 1. Figure 1A corresponds to samples labelled with T-Red310, and Figure 1B corresponds to the same gel in silver stain. Using 1-DE a complex banding pattern was obtained for all saliva samples. The pattern before transport was variable between individuals. Nevertheless, an increase in the number and intensity of bands can be observed after transport.

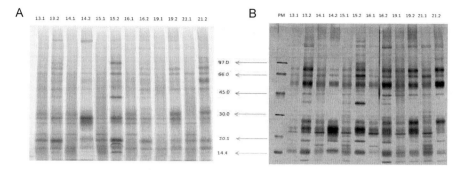

Sample #.1: sample collected before transport
Sample #.2: sample collected after transport

Figure 1. 1-DE, from saliva samples of pigs before and after a long transport. (A) T-Red 310 stain. (B) Silver stain.

Likewise, 2-DE separation (Figure 2) also showed differences in the number and intensity of several spots when comparing samples obtained before (Figure 1A) and after transport (Figure 2B). In this case, the image corresponds to samples labelled with T-Red310.

Comparison between staining methods

Silver staining gives more intensely stained patterns than T-Red, in 1-DE as well as in 2-DE (Figure 1), giving the optical impression of higher contrast. Nevertheless, T-Red has a larger linear range than silver staining, so is less prone to saturation effects (which is beneficial for quantification). It was also noticed that some spots or bands were not equally well detectable in both methods, possibly due to different staining mechanisms.

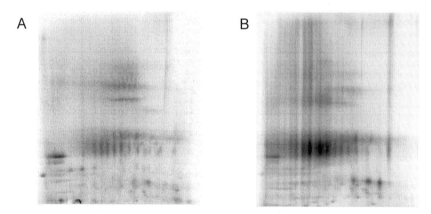

Figure 2. 2-DE pattern from saliva samples of pigs before (A) and after (B) a long transport; T-Red310 stain.

Some bands were putatively identified by homology with the saliva protein pattern described in the literature (Gutiérrez *et al.*, 2013); for example: albumin, haptoglobin, and two bands in the region of the IgL chain and lipocalin. Since most probably the protein pattern before transport was affected by food intake, these samples were not considered for further analysis and we focused on the comparison between samples after transport. Assuming the preliminary identification of the protein bands, changes after short (no stress) or long (stress) transport were quantified by densitometry and results are shown in Table 1.

Table 1. Average and S.D. of densitometry values.

Protein	Condition	Mean	S.D.
Albumin	No stress	11,860	13,796
	Stress	16,090	10,293
Haptoglobin	No stress	7,230	6,003
	Stress	13,458	9,092
IgL chain	No stress	126,997	73,874
	Stress	100,476	80,264
Lipocalin	No stress	13,986	9,500
	Stress	8,392	10,513

Conclusions

Differences in saliva protein composition between samples from pigs after a short or long transport have been observed. Nevertheless, the identity of these protein bands needs to be confirmed by mass spectrometric analyses or Western blots.

References

Gutiérrez, A.M., Miller, I., Hummel, K., Nöbauer, K., Martínez-Subiela, S., Razzazi-Fazeli, E., Gemeiner, M., Cerón, J.J. (2011). Proteomic analysis of porcine saliva. The Veterinary Journal 187, 356-362.
Gutiérrez, A.M., Nöbauer, K., Soler, L., Razzazi-Fazeli, E., Gemeiner, M., Cerón, J.J., Miller, I. (2013). Detection of potential markers for systemic disease in saliva of pigs by proteomics: A pilot study. Veterinary Immunology and Immunopathology 151, 73-82.
Obayashi, K. (2013). Salivary mental stress proteins. Clinica Chimica Acta 425, 196-201

Identification of welfare biomarkers in pigs using peripheral blood mononuclear cells and differential gel expression

Daniel Valent, Laura Arroyo, Lourdes Criado, Francesca Dell'Orco and Anna Bassols*
Departament de Bioquímica i Biologia Molecular, Facultat de Veterinària. Universitat, Autònoma de Barcelona, Spain; danivalent457@gmail.com

Introduction

Animal welfare is a topic of high interest due to its implications for animal health and because of economic consequences. The objective of this work was to identify protein markers for welfare assessment. Peripheral mononuclear cells (PBMCs) are a sample of interest since they can be easily obtained and do not contain high abundant proteins as plasma. To achieve this objective we collaborated with the Animal Welfare Unit of IRTA (Institut de Recerca i Tecnologia Agroalimentària) who was responsible for animal care and management.

Sus scrofa specie was chosen as an animal model due to high interest to livestock farming industry and for the easiness of stable in an experimental farm. Only females pigs were used in our experiment to avoid the sex variability with the purpose of simplify DIGE analysis.

Materials and methods

Two groups of pigs (3-month old, n=15) were subjected to different management conditions. One group was kept under commercial farm conditions ('non-treated', NT) and the other one was subjected to a close human-animal relationship ('treated', T), with daily personal care of the animals by the same, well trained person. Serum and PBMCs were collected at 0, 1 and 2 month. Biochemical nutritional parameters, acute phase proteins, antioxidant enzymes and cortisol were determined in serum to control the health status of the animals. PBMCs were purified using ficoll gradient tubes (BD Vacutainer® CPT™) and stored at -80 °C.

PBMC cell pellets were lysed using a lysis buffer (urea 7 M, thiourea 2 M, CHAPS 4%, proteases inhibitors and pH 8) and sonicated to obtain the protein extract. Then proteins were precipitated with Clean Up Kit (GE Healthcare™) and resuspended in urea-thiourea buffer. Finally protein concentrations were quantified using RC/DC Protein Assay (BioRad Kit).

Differences in protein expression between groups were analyzed by Two dimensional electrophoresis-DIGE using 4 animals from each group, at time 0 (3 month-old) and 2 (5 month-old).

After DIGE, the relevant spots were picked using a Spotspicker machine and digested with trypsin and analysed on an LTQ Obitrap Velos (Thermo Fisher Scientific, Waltman, MA).

Results

When comparing differences in protein expression between t=0 and t=2 month, 305 differential spots were identified in non-treated animals and 153 in the treated group ($P<0.05$ and ≥ 1.5 fold-change). Eighty-five spots overlapped between NT and T groups, whereas 220 spots were differentially expressed only in the NT group and 68 spots in the T group.

A total of 64 protein spots were selected on the basis of fold-change and p-value and subsequently identified by LC-MS/MS (Figure 1).The selection of spots for sequencing maintains the different expression relation between NT and T conditions (Table 1).

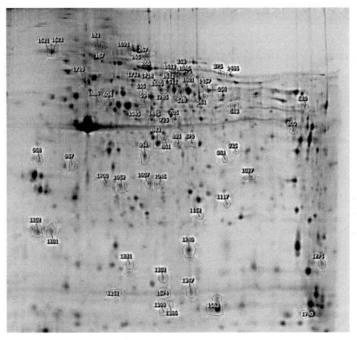

Figure 1. Proteome pattern of PBMC cell lysate. Spots selected for identifications have been surrounded with a line.

Table 1. Number of spots picked that display differences between t=0 and t=2 in non-treated (NT) and treated (T) groups.

	Protein expression					Total
	NT>T	T>NT	Only NT	Only T	No changes	
No. of spots	43	11	5	0	5	64

Table 2 shows a partial list of the proteins identified in both animal groups using Pig and Other Mammalia databases from Uniprot (Table 2).

Finally the list of the 64 identified proteins was introduced in Panther Classification System to represent a pie chart graphic and classify the proteins pursuant to molecular function and cell distribution (Figure 2).

Conclusions

Differences in protein expression between t=0 and t=2 are probably due to age since the animals were still young and growing. Nevertheless, a larger number of changes were found in non-treated animals than in animals subjected to a close relationship to humans, suggesting that some of the changes were due to management conditions and not only to age.

Table 2. Partial list of spots identified showing fold change differences respect non-treated (NT) and treated (T) animals.

Spot no.	Protein name	Fold change	
		NT t=0/NT t=2	T t=0/T t=2
353	Far upstream element-binding protein 2	3.4	-
561	tRNA-splicing ligase RtcB homolog	2.0	-
879	PDZ and LIM domain protein 1	1.5	2.0
1240	Calcium-binding protein A9	11.8	4.3
1607	High mobility group protein B1	1.9	2.0

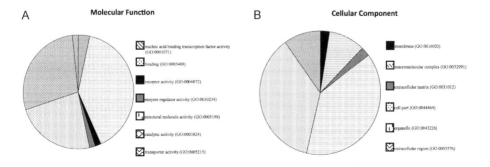

Figure 2. Pie chart diagrams showing the different molecular functions and cell distribution of the 64 proteins that change their expression throughout the time.

Proteins found in this experiment mainly belong to the following GO classes regarding molecular function: binding proteins (40%) mostly related to protein-protein interaction, structural proteins involved in cell cytoskeleton (23%) and proteins with catalytic activity (28%) specifically hydrolase activity.

Furthermore these proteins are mostly located to cytoplasm (39%), actin cytoskeleton and mitochondria (37%).

In conclusion, protein expression in PBMCs can be modified even by relatively mild changes in pig management conditions. Although still preliminary, the present results can give insight into the physiological mechanisms of adaptation to environmental conditions and help to identify protein markers for welfare assessment.

References

Danielsen, M., Pedersen, L. J., Bendixen, E., 2011. An *in vivo* characterization of colostrum protein uptake in porcine gut during early lactation. Journal of Proteomics 74, 101-109.

Muthukumar, S., Rajkumar, R., Rajesh, D., Saibaba, G., Liao, C., Archunan, G., Padmanabhan, P., and Gulyas, B., 2014. Exploration of salivary proteins in buffalo: an approach to find marker proteins for estrus. FASEB Journal 28, 000-000.

Ramirez-Boo, M., Garrido, J. J., Ogueta, S., Calvete, J. J., Gómez-Díaz, C. and Moreno, A., 2006. Analysis of porcine peripheral blood mononuclear cells proteome by 2-DE and MS: Analytical and biological variability in the protein expression level and protein identification. Proteomics 6, S215-S225.

A search for protein biomarkers links olfactory signal transduction to social immunity

M. Marta Guarna[1]*, Andony P. Melathopoulos[2,§], Elizabeth Huxter[3], Immacolata Iovinella[4], Robert Parker[1,‡], Nikolay Stoynov[1], Amy Tam[1], Kyung-Mee Moon[1], Queenie W. T. Chan[1], Paolo Pelosi[4], Rick White[5], Stephen F. Pernal[2] and Leonard J. Foster[1]

[1]Department of Biochemistry & Molecular Biology, Centre for High-Throughput Biology, University of British Columbia, Vancouver, BC, V6T 1Z4, Canada; gmarta@outlook.com
[2]Beaverlodge Research Farm, Agriculture & Agri-Food Canada, Beaverlodge, AB, T0H 0C0, Canada
[3]Kettle Valley Queens, Grand Forks, BC, Canada
[4]Department of Chemistry, University of Pisa, Italy
[5]Department of Statistics, University of British Columbia, Vancouver, BC, V6T 1Z4, Canada
Current addresses:
§ Dalhousie University, Halifax, NS, Canada
‡ Macquarie University, Sydney, NSW, Australia

Background

The Western honey bee (*Apis mellifera* L.) is a critical component of human agriculture through its pollination activities. For years, beekeepers have controlled deadly pathogens such as *Paenibacillus larvae*, *Nosema* spp. and *Varroa destructor* with antibiotics and pesticides but widespread chemical resistance is appearing and most beekeepers would prefer to eliminate or reduce the use of in-hive chemicals. While such treatments are likely to still be needed, an alternate management strategy is to identify and select bees with heritable traits that allow them to resist mites and diseases. Breeding such bees is difficult as the tests involved to identify disease-resistance are complicated, time-consuming, expensive and can misidentify desirable genotypes. Additionally, we do not yet fully understand the mechanisms behind social immunity. Here we have set out to discover the molecular mechanism behind hygienic behaviour (HB), a trait known to confer disease resistance in bees.

Results

We first confirm that honey bees can be selectively bred for hygienic behaviour and then, over three years, two geographically distinct sites and several hundred bee colonies, we correlated protein expression with measured behaviours in each colony. By correlating the expression patterns of individual proteins with HB scores, we identified seven putative biomarkers of HB that survived stringent control for multiple hypothesis testing. Intriguingly, these proteins were all involved in semiochemical sensing (odorant binding proteins), nerve signal transmission or signal decay, the series of events required in responding to an olfactory signal from dead or diseased larvae. We then use recombinant versions of two odorant-binding proteins to identify the classes of ligands that these proteins might be helping bees detect.

Conclusions

Our data suggest that neurosensory detection of odours emitted by dead or diseased larvae is the likely mechanism behind a complex and important social immunity behaviour that allows bees to co-exist with pathogens.

Bacterial competition for food safety in dairy products

Isabella Alloggio[1]*, Mylène Boulay[2], Vincent Juillard[2], Véronique Monnet[2,3], Alessio Soggiu[1], Cristian Piras[1], Luigi Bonizzi[1] and Paola Roncada[1,4]

[1]DIVET, Dipartimento di Scienze Veterinarie e Sanità Pubblica, Università degli Studi di Milano, Italy; isabella.alloggio@unimi.it

[2]INRA, MICALIS, UMR1317, 78350 Jouy en Josas, France

[3]INRA, PAPPSO, 78350 Jouy en Josas, France

[4]Istituto Sperimentale Italiano L. Spallanzani, Milano, Italy

Introduction

Listeria monocytogenes is an ubiquitous pathogen responsible for several outbreaks/foodborne diseases. The mortality rate is close to 24% mainly in immunocompromised persons (Farber and Peterkin, 1991). In the spreading of this pathology, milk and dairy products are a key reservoir for this pathogen (Greenwood *et al.*, 1991). This represents a serious burden if considering the possibility of human infection and the financial losses due to the strict rules for food export. Food processing is one of the major steps that could be linked to *L. monocytogens* contamination (Northolt *et al.*, 1988) that is probably due to the presence of *L. monocytogens* after the first post-pasteurization process of milk (Kozak *et al.*, 1996). Inhibition of *L. monocytogens* growth through the competition of other bacteria could represent a solution to this problem. In particular the production of bacteriocins by some species of *Lactococcus* could play a key role in pathogens growth inhibition (Stecchini *et al.*, 1995). As well as, the study of the putative production of short genes coding peptides of *L. monocytogens* could represent an important point, especially if considering their putative role in bacterial gene regulation. The objectives of this work are the bioinformatic study of the genomes of Listeria monocytogenes strains, in order to predict putative short genes coding peptides, potentially involved in gene regulation and the evaluation of extracellular peptidome analysis of separated lactoccoccus and listeria cultures and of these bacteria growing in competition conditions.

Materials and methods

Bacterial strains

Two strains of *Lactococcus lactis* (ATCC 11454, IL1403) and two strains of *L. monocytogens* (ATCC 19115, EDGe) have been used in the present study. They were stored at -80 °C in M17 broth containing glycerol.

Culture conditions

Strains of both bacteria were precultured at 30 °C in CDM (chemically-defined medium) (Letort and Juillard, 2001) and incubated overnight. 1% of *L. lactis* preculture was inoculated in 50 ml of MCD for *L. lactis* monoculture. 1% of *L. monocytogenes* preculture was inoculated in 50 ml of MCD for *Listeria* monoculture. 0.5% of *Lactococcus* preculture and 0.5% *Listeria* preculture were inoculated in 50 ml of CDM for coculture. Three described cultures were incubated at 30 °C. Bacterial growth was estimated by measuring optical density at 600 nm (OD600) to the end of the exponential phase of growth.

Supernatant preparation

Cultures were centrifuged (5000 rpm, 10 min, 4 °C) and supernatants were recovered. Supernatant was filtered using PDVF membrane 0.22 µm. Samples concentration has been performed using SPE (Strata X, 200 mg, 0.3 ml/min) and ultrafiltration using 3 kDa cut-off membranes (Amicon).

HPLC injection in neutre condition

200 µl of each sample was separated by HPLC system equipped with an Acclaim column C18, 3µm, 2.1×150mm, 300 A°. Mobile phase A was ammonium formate 20 mM and mobile phase B was ammonium formate 20 mM in 80% acetonitrile. Peptides were separated with a gradient of 5-35% mobile phase B over 20 min, followed by a plateau of 35% mobile phase B. Fractions were collected from 0 to 35 min. Fractions were dried by evaopartion (Speed Vac) then resuspended in 30 µl of TFA 0.08%/ACN 2%

LC-MS/MS analysis

LC-MS/MS analysis was performed on the PAPPSO platform (INRA, Jouy-en-Josas, France). An Ultimate 3000 LC system (Dionex) was connected to to a linear ion trap mass spectrometer (LTQ, Thermo Fisher) by a nanoelectrospray interface to conduct the separation, ionization and fragmentation of peptides, respectively. 5 microliters of each sample were loaded at a flow rate of 20 µl/min onto a precolumn (Pepmap C18; 0.3×5 mm, 100 Å, 5 µm; Dionex). After 4 min, the precolumn was connected with the separating nanocolumn Pepmap C18 (0.075 by 15 cm, 100Å, 3 µm), and the linear gradient was started from 2 to 36% of buffer B (0.1% formic acid, 80% ACN) in buffer A (0.1% formic acid, 2% ACN) at 300 nl/min over 50 min. Ionization was performed on liquid junction with a spray voltage of 1.3 kV applied to an uncoated capillary probe (PicoTip EMITER 10-µm tip inner diameter; New Objective). Peptides ions were automatically analyzed by the data-dependent method as follows: full MS scan (m/z 300 to 1,600) on Orbitrap analyzer and MS/MS on the four most abundant precursor on the LTQ linear ion trap. Data obtained in the instrument-specific data format (.RAW) were converted to mzxml files for further data analysis using a conversion software program (MSConvertGUI). Peptidomic data were analyzed by X!Tandem Pipeline software.

Bioinformatic analysis

The genomic sequence of strains has been analyzed for the presence of short genes at the MIGALE platform (INRA, Jouy-en-Josas, France) using the BactGeneShow program. A gene containing from

48 to 183 bases (peptide from 15 to 60 amino acids) is considered as a short gene (artificial cut off), genes containing more than 183 bases are considered as 'normal' genes. The threshold that has been used is mainly based on removal of predictions related to genes shorter than 48 bases.

Three steps are fundamental for the construction of the database used for the peptides identification:
1. Extraction of the regions corresponding to coding sequences.
2. Reversion of the nucleotidic sequences that are located on the reverse DNA strand.
3. Conversion from nucleotides to amino acids.

All these steps are done using bio-informatic scripts that are enclosed in the EMBOSS package.

Results

In Figure 1, growth curves of monoculture and coculture of *L. monocytogenes* and *L. lactis* with different strains are shown.

Database searching, performed by X! Tandem Pipeline, allowed the identification of peptides that accumulates in the medium during the growth of the strains. About 957 peptides were identified for the *L. monocytogens* ATCC 19115 monoculture, 2350 for *L. lactis* ATCC 11454 and 1440 for the mixed culture. 957 petides derive from 115 proteins for the monoculture of *L. monocytogenes* ATCC 19115; 2350 peptides from 110 proteins for the monoculture of *L. lactis* ATCC 11454 and 1440 peptides derive from 115 proteins identified in mixed culture (Figure 2A).

Figure 2B shows a representative distribution of proteins identified in monoculture of *L. monocytogenes* EGDe, in monoculture of *L. lactis* IL 1403 and in co-culture (*Listeria-Lactococcus*). 984 came from the degradation (by the bacteria) of 100 proteins for the monoculture of *L. monocytogenes* EDGE, among these 30 were present also in co-culture. Moreover, 4 proteins were expressed only in co-culture condition. 1741 peptides derive from 122 proteins identified

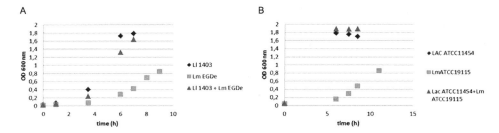

*Figure 1. (A) Growth curves of monoculture (*Listeria monocytogenes *EGDe,* Lactococcus lactis *IL1403) and co-culture (*L. monocytogenes *EGDe –L.* lactis *IL 1403). (B) Growth curves of monoculture (*Listeria *ATCC 19115, L. lactis* ATCC 11454) and co-culture (*L. monocytogenes* ATCC 19115-L.* lactis *ATCC 11454).*

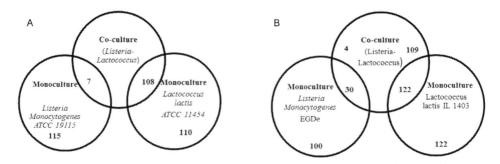

Figure 2. (A) Representative distribution of proteins identified in Monoculture (Listeria monocytogenes ATCC 19115, Lactococcus lactis ATCC 11454) and in co-culture (Listeria-Lactococcus). (B) Representative distribution of proteins identified in Monoculture (L. monocytogenes EDGe, L. lactis IL 1403) and in co-culture (Listeria-Lactococcus).

in monoculture of *L. lactis* IL1403 and also in co-culture. 9 other proteins were expressed only in co-culture condition. 2587 peptides came from 165 proteins identified in coculture condition.

This work has shown that specific proteins are degradated during the co-culture and we are currently investigating to elucidate the mechanism involved.

Acknowledgements

This work was performed during a COST ACTION FA1002 FAP Short Term Scientific Mission at INRA, UMR1319 MICALIS, Jouy-en-Josas(FR). This project was supported by Ministry of Health CCM Project: Milano EXPO 2015 (LB)

References

Farber, J. and Peterkin, P., 1991. Listeria monocytogenes, a food-borne pathogen. Microbiological reviews 55: 476.

Greenwood, M., Roberts, D. and Burden, P., 1991. The occurrence of Listeria species in milk and dairy products: a national survey in England and Wales. International Journal of Food Microbiology 12: 197.

Kozak, J., Balmer, T., Byrne, R. and Fisher, K., 1996. Prevalence of Listeria monocytogenes in foods: Incidence in dairy products. Food Control 7: 215-221.

Letort, C. and Juillard, V., 2001. Development of a minimal chemically defined medium for the exponential growth of Streptococcus thermophilus. Journal of applied microbiology 91: 1023-1029.

Northolt, M., Beckers, H., Vecht, U., Toepoel, L., Soentoro, P. and Wisselink, H., 1988. Listeria monocytogenes: heat resistance and behaviour during storage of milk and whey and making of Dutch types of cheese. Netherlands Milk and Dairy Journal (Netherlands).

Stecchini, M.L., Aquili, V. and Sarais, I., 1995. Behavior of Listeria monocytogenes in Mozzarella cheese in presence of Lactococcus lactis. International Journal of Food Microbiology 25: 301-310.

Mule duck 'foie gras' show different metabolic states according to their quality phenotypes by using a proteomic approach: comparison of 2 statistical methods

Yoannah François[1,2,3,4], Christel Marie-Etancelin[2,3,4], Alain Vignal[2,3,4], Didier Viala[5], Stéphane Davail[1] and Caroline Molette[2,3,4]*

[1]*Université de Pau et des Pays de l'Adour, UMR5254, 40 004 Mont de Marsan Cedex, France*
[2]*INRA, UMR1388 Génétique, Physiologie et Systèmes d'Elevage (GenPhySE), 31326 Castanet-Tolosan, France*
[3]*Université de Toulouse, INPT ENSAT, UMR1388 GenPhySE, 31326 Castanet-Tolosan, France; molette@ensat.fr*
[4]*Université de Toulouse, INPT ENVT, UMR1388 GenPhySE, 31076 Toulouse, France*
[5]*INRA, U0370 PFEMcp, Theix, 63122 Saint Genès Champanelle, France*

Introduction

France is the main producer of 'foie gras' (72% of world production) and 97% of this production comes from ducks. The 'foie gras' is rich in lipids and as this high fat content is responsible for its organoleptic qualities, too much fat loss during the cooking process is a major problem. Recently, Théron *et al.* (2011) used a proteomics approach to explore the influence of protein composition on fat loss variability, for a limited range of liver weight and of liver lipid contents. Finally, Kileh-Wais *et al.* (2013) identified QTL related to several 'foie gras' quality traits such as liver weight, lipid rate, protein rate, melting rate, color lightness, collagen rate, although the genes underlying these QTL still have to be discovered.

In this context, the aim of the present study is to decipher mechanisms involved in the quality of 'foie gras' by means of a proteomics approach using 2D gel electrophoresis. To deepen our understanding of the mechanism underlying 'foie gras' quality variability, complex traits are measured, namely the proportion of protein content in crude liver (crude liver protein content) and in dry liver (dry liver protein content) in addition to the classical liver weight and melting rate traits. The proteomics approach allowed the identification of proteins present at different levels of abundance in the samples studied, in relation to the melting rate level alone or to the other traits by using 2 different statistical procedures. Thereafter, biological pathway analysis gives some information on the metabolic state of the livers in relation to their phenotypes and allows us to better understand the biological mechanisms underlying the variations of 'foie gras' quality.

Material and methods

The animals are male mule ducks (n=294). They were bred, over-fed and slaughtered at the experimental unit dedicated to waterfowl (UEPFG, Benquet, France).

The 2DE analysis (n=294) was performed according to Théron *et al.* (2011). Briefly, the proteins were solubilized in a standard buffer (Tris HCl 1.5 M pH 8). Bradford method was used for the protein assay. After an isoelectric focusing (pH range 5-8), the strip is deposited on the surface of SDS-PAGE 12% acrylamide. The gels are stained with brilliant blue G250 and scanned using the scanner ImageScanner III. The image analysis software was performed with Samespots software® (TotalLab Ltd, Newcastle-upon-Tyne, UK). The raw data were first corrected for the fixed effects: the zootechnical ones (crammers and hatching batching effects) and the technical ones (gels quality and batches of each dimension electrophoresis). Then, the proteomic data were analysed according to the 4 quality traits, either using a variance analysis (GLM procedure of SAS®) with each quality trait split in 4 quartiles, or using a linear regression (REG procedure of SAS®). In both cases, the p-values were corrected for multiple tests with the Benjamini-Hocheberg procedure.

The spots of interest, with different abundances according to the levels of quality parameters (melting rate, protein content and liver weight), were identified by mass spectrometry in proteomics platform of INRA Theix. For Nano-LC-MS/MS Analysis, peptides mixtures were analyzed by online nanoflow liquid chromatography using the Ultimate 3000 RSLC (Dionex, Voisins le Bretonneux, France). For raw data processing, Thermo Proteome Discoverer 1.4 (v:1.4.0.288) was used with MASCOT (v:2.3) for database search (www.matrixscience.com). UniP_tax_Aves (150722 sequences) was used as database for protein identification.

Results and discussion

Quartiles of the 3 phenotypic traits are represented in Table 1. Whatever the quality trait, 45 spots (30 identified) were found to be differentially expressed by the regression (REG) procedure whereas 35 (23 identified) spots were pointed out with the GLM procedure. Among the 30 identified spots with the REG method, 16 are significantly affected by only one out of the 4 quality traits, 5 are significantly affected by 2 or 3 traits and 9 spots (ALB, FKBP4, ENO1, ANXA5 FASN, PRDX6 and VCP) are common to the 4 traits. Among the 23 protein spots pointed out by the ANOVA, 15 had a

Table 1. Repartition in quartiles of phenotypic values.

	Quartile 1	Quartile 2	Quartile 3	Quartile 4
LW (g)	394±75	531±27	621±25	731±59
MR (%)	18.7±5.1	34.8±5	46.5±2.5	55.2±3.9
LprotCc (%)	6.2±0.4	7.1±0.2	7.9±0.2	9.4±1.0
LprotDc (%)	9.0±0.7	10.7±0.4	12.4±0.5	16.3±2.9

LW=Liver Weight, MR=Melting Rate expressed as a percentage of the liver quantity before cooking, LprotCc=Crude Liver protein content expressed as a percentage of the fresh liver weight, LprotDc=Dry liver protein content expressed as a percentage of the dry liver weight.

variation in abundance following a single trait only (LW, MR or LprotDc), whereas the 7 remaining spots were differentially present following 2 traits simultaneously, only 1 spot (FASN) following 3 traits and none are common to the 4 traits. Among the 23 and 30 spots differentially expressed and identified according to the GLM and REG procedures respectively, 20 are common to both methods, 4 specific to the GLM and 10 to the REG. Whatever the statistical analysis used, the proteins showing variation in our dataset represent several biological pathways such as glycolysis, lipid metabolism, oxidative stress, transport, catabolism and anabolism processes.

The comparison of two approaches ANOVA and REG reveal specific proteins. For example, two heat shock proteins (HSPD1 and HSPA5) disappear when using REG procedure because they did not evolve linearly with the trait of interest (HSP expression in the last quartile with MR is much lower than expected under the linear hypothesis). HSPs are synthesized proteins following stress (Heat Shock Factor) in order to respond to it (Niforou *et al.*, 2014). HSPA5 is of the family of HSP70, a cytosolic protein can translocate into the nucleus during a 'heat shock'. HSPD1 of the HSP60 family is a mitochondrial chaperone protein that helps for the conformation of proteins imported into the mitochondria. Stimulation of HSP is successive to stress; their non-linearity with respect to the MR suggests that the stimulation is due to a threshold effect and therefore is probably more timely than gradual. Furthermore, their synthesis may be just the time to stimulate and support the synthesis of other overexpressed proteins of cytoprotection in the last quartile of the weight of liver (HYOU1) or the MR (PRDX6). Four proteins were only identified using REG procedure: PARK7 and ANXA2, affected by LW are cytoprotecting proteins such as PRDX6 and HYOU1, PFN2, affected by MR, is implicated in cell motility and TCP1, affected by LprotDc, is a chaperonin protein. These proteins bring precisions on a link between pathways and phenotypes. Under linear hypothesis, a regression seems more efficient than variance analysis in identifying such links.

When looking at the protein profiles (Figure 1), whose variation in protein abundance levels concerns 2 or more traits, a remarkable opposition appeared between the trend of proteins expression with LW and MR levels on the one hand and LprotDc level on the other hand. If protein expression levels increased with LW (FKBP4, C11orf54, PRDX6, ENO1), it also increased with MR or decreased with LprotDc. Inversely, if protein expression levels decreased with MR (APOA1, ALB and FASN), they increased for LprotDc, with the notable exception of VCP. Proteins which expressions decrease with MR or LW and/or increase with LprotDc represent pathways such as synthesis process (LOC100545435, FASN), anabolism process (DLAT, ME1), transcription regulation (PHB) and transport (APOA1, ALB). On the contrary, proteins which expressions increase with MR or LW and/or decrease with LprotDc represent biological pathways such as response to stress (HYOU, HSPA5, HSPD1, and PRDX6), transport (FABP1), glycolysis (ENO1), and catabolism process.

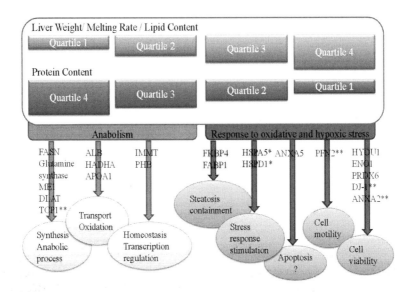

Figure 1. Summary diagram of the different liver protein profiles (from A to F) according to their phenotype. FASN=Fatty acid synthase, ME1=Malic enzyme, DLAT=Dihydrolipoamide S-acetyltransferase, ALB=Albumin, HADHA=Trifunctional enzyme subunit alpha, APOA1=Apolipoprotein A1, IMMT=Mitochondrial inner membrane protein-like, PHB=Prohibitin, PRDX6=Peroxiredoxin 6, HSPA5=70kDa Heat shock protein 5, HSPD1=60 kDa Heat shock protein, FKBP4=FK506 −Binding Protein 4, FABP1=Fatty acid binding protein 1, ANXA5=Annexin A5, HYOU1=Hypoxia up-regulated protein 1, ENO1=α-enolase, TCP1: T-complex protein 1; PFN2: Profilin 2; ANXA2: Annexin A2; DJ-1: PARK7 Protein names with stars were only identified by GLM/ANOVA (1 star) or by REG (2 stars).

References

Kileh-Wais, M., Elsen, J.M., Vignal, A., Feves, K., Vignoles, F., Fernandez, X., Manse, H., Davail, S., André, J.M., Bastianelli, D., Bonnal, L., Filangi, O., Baéza, E., Guéméné, D., Genêt, C., Bernadet, M.D., Dubos, F. and Marie-Etancelin, C., 2013. Detection of QTL controlling metabolism, meat quality, and liver quality traits of the overfed interspecific hybrid mule duck. J. Anim. Sci. 91, 588-604.

Niforou, K., Cheimonidou, C. and Trougakos, I.P., 2014. Molecular chaperones and proteostasis regulation during redox imbalance. Redox Biology 2, 323-332.

Theron, L., Fernandez, X., Marty-Gasset, N., Pichereaux, C., Rossignol, M., Chambon, C., Viala, D., Astruc, T. and Molette, C. 2011. Identification by proteomic analysis of early post-mortem markers involved in the variability in fat loss during cooking of mule duck 'foie gras'. J. Agric. Food Chem. 59, 12617-12628.

Differential peptidomics of raw and pasteurized sheep milk cheese

M. Filippa Addis[1]*, Salvatore Pisanu[1], Daniela Pagnozzi[1], Massimo Pes[2], Antonio Pirisi[2], Roberto Anedda[1] and Sergio Uzzau[1]

[1]*Porto Conte Ricerche, SP 55 Porto Conte/Capo Caccia, Loc. Tramariglio, 07041 Alghero, Sassari, Italy; addis@portocontericerche.it*

[2]*AGRIS Sardegna, Department of Animal Science, Loc. Bonassai, 07040 Olmedo, Sassari, Italy*

Introduction

During cheese ripening, milk proteins undergo progressive degradation due to residual rennet activity, endogenous proteolytic enzymes from milk, microbial enzymes, and exogenous proteinases or peptidases added to improve cheese maturation (McSweeney, 2004). Due to these different contributions, production variables can lead to changes in the final peptide profile of ripened cheese. One of these variables is the treatment of milk with high temperatures, aimed to eliminate pathogens and to improve cheese keeping quality. However, exposure to heat can also produce changes in milk proteins, including denaturation, formation of complexes, modification of amino acid side chains and post-translational modifications; in addition, it can disrupt the endogenous microbial flora, and cause denaturation and chemical modification of indigenous proteolytic enzymes and kinases (Gaya *et al.*, 2005; Raynal-Ljutovac *et al.*, 2007; Mendia *et al.*, 2000; Singh, 2004). All these events can impact the final peptide profile of ripened cheese and, adding to the more relevant sensory and textural attributes, bioactivity features can also be affected (Korhonen *et al.*, 1998). The bioactive potential of a cheese is due to different molecules including bioactive peptides, 2 to 20-mer aminoacid sequences that possess immunomodulatory, antihypertensive, antioxidative, antimicrobial, anticancer, and opioid agonist activities (Korhonen, 2009; Nagpal *et al.*, 2011). Sheep milk cheese, or Pecorino, is one of the most relevant productions in several Mediterranean areas, including the Island of Sardinia, where one of the typical PDO productions, Fiore Sardo, is made with raw milk in agreement with the production specifications. Other traditional cheeses, including Pecorino di Filiano, Ossau-Iraty, Roquefort, Roncal and Idiazabal are also made from raw sheep milk. To investigate if and how pasteurization impacts on total and bioactive peptides abundances, sheep cheese was prepared with raw or pasteurized milk in controlled conditions, ripened for 9 months, and subjected to extraction and characterization of peptides by liquid chromatography-tandem mass spectrometry (LC-MS/MS); then, peptides were evaluated for relative abundance, originating proteolytic enzymes, and bioactive potential.

Materials and methods

Cheese was prepared in a pilot plant by processing, in three different days, 50 l of raw milk and pasteurized milk obtained from the same batch of 100 l split into two aliquots, for a total of six preparations. Pasteurization was carried out in an infrared tubular exchanger (model Stoutz-

Actinator, ACTINI GROUP, France), by heating milk at 72 °C for 15 s followed by quick cooling at 38 °C. Freeze-dried mixed starter culture and liquid calf rennet were added and, after clotting, the coagulum was cut, moulded, and then left at 32 °C for 5 h before keeping in saturated brine for 24 h. Ripening was carried out for nine months at 12 °C and 85% relative humidity. Water-soluble extracts (WSEs) were prepared from minced cheese based on the protocol by Rizzello *et al.* (2005). After lyophilization, 10 g of cheese resuspended in 50 mM sodium phosphate buffer, pH 7.0, were homogenized by using a TissueLyser II (Qiagen, Hilden, Germany) and incubating under gentle agitation at 40 °C. Filtered supernatants were adjusted to pH 4.6 to precipitate caseins, centrifuged, and quantified. WSEs underwent differential solubilization (DS) as described previously (Biosa *et al.*, 2011), and then quantitation by the BCA assay (Thermo Scientific, Rockford, USA). Peptides were dissolved in 0.2% formic acid and analyzed on a Q-TOF hybrid mass spectrometer equipped with a nano lock Z-spray source and coupled on-line with a capillary chromatography system CapLC (Waters, Manchester, UK), as described before (Tanca *et al.*, 2011). Peptide abundance was assessed using the spectral counts (SpC) of each peptide for calculating the normalized spectral abundance factor (NSAF), as described previously (Old *et al.*, 2005). NSAFs were subjected to principal component analysis (PCA) using SIMCA-P software (version 12.0.1, Umetrics, Umea, Sweden). SpC log ratios (R_{SC}, R vs P) were calculated according to Old *et al.* (2005). Differential peptides were evaluated with the beta-binomial test with FDR adjustment (Pham *et al.*, 2010), by considering only peptides with $R_{SC} \geq \pm 1.5$ and $P \leq 0.05$ as significant. EnzymePredictor (http://bioware.ucd.ie/~enzpred/Enzpred.php) (Vijayakumar *et al.*, 2012) and BIOPEP (http://www.uwm.edu.pl/biochemia/index.php/pl/biopep) (Minkiewicz *et al.*, 2008), were used for estimating proteolytic cleavages and bioactivities, respectively.

Results and discussion

A total of 187 different peptides were identified from raw milk cheese (R) and pasteurized milk cheese (P). Upon comparison of the identity and abundance of all peptides identified, ripened R and P showed differences both in terms of peptide identities and abundances. Concerning peptide sequences, 55 peptides were unique, 32 being present only in R and 23 only in P. By applying statistical filters and upon label-free quantification, 58 peptides were also significantly different in abundance between R and P. In general, moreover, proteolysis was less intense and more homogeneous in P than in R. To investigate intensity and nature of the proteolytic processes, the 187 peptide sequences were processed with Enzyme Predictor, a web-based software that enables to predict the enzymes responsible for peptide generation (Vijayakumar *et al.*, 2012). The software predicted a lower number of cuts for P when compared to R, for almost all the enzymes considered, suggesting that inactivation of indigenous proteolytic enzymes can play a role in the lower level of proteolysis seen in P, but also that changes induced by heat on milk proteins can ultimately influence their degradation by reducing susceptibility to proteases, since a lower number of cuts was predicted for almost enzymes. In order to investigate potential bioactivity of ripened R and P cheese, all sequences were processed with the BIOPEP application (Minkiewicz *et al.*, 2008). Out of 187 sequences, 37 were associated to a bioactive function, including immunomodulation, ACE-inhibition, and antibacterial, antioxidative, anticancer and opioid-agonist activities. According to NSAF values, immunomodulating and ACE-inhibitor activities were significantly higher in R ($P \leq 0.05$), while opioid activity was higher in P. All other biological activities were comparable. Therefore, in addition to sensory and textural changes, pasteurization of milk for cheese production can also influence bioactivity of the ripened product, with a potential impact on consumer health. In view of these interesting results, we are currently evaluating a more profound characterization approach based on peptide enrichment by FPLC, peptide characterization by LTQ-Orbitrap Velos MS, estimation of ACE-inhibition by HPLC, and assessment of antimicrobial activity by measurement of MICs on microbial cultures.

Conclusion

The peptide profile of ripened cheese made from raw or pasteurized sheep milk was different both in terms of sequences identified and their abundances. Likely, exposure of milk to high temperatures impacts on suitability of its proteins to enzymatic degradation as well as on indigenous flora or enzymes that act on the protein substrates. In addition to sensory and textural features, the bioactive potential of the final product also appears to be influenced. The identification of peptide molecules with beneficial properties in sheep cheeses made from raw milk can provide important opportunities for valorization and preservation of traditional sheep cheeses. The characterization of peptide profiles in sheep cheese by proteomics can also provide useful hints for assessing the impact of production process variables on the final, ripened product.

References

Biosa, G., Addis, M. F., Tanca, A., Pisanu, S., Roggio, T., Uzzau, S. and Pagnozzi, D. 2011. Comparison of blood serum peptide enrichment methods by Tricine SDS-PAGE and mass spectrometry. Journal of Proteomics 75, 93-99.

Gaya, P., Sánchez, C., Nuñez, M. and Fernández-García, E. 2005. Proteolysis during ripening of Manchego cheese made from raw or pasteurized ewes' milk. Seasonal variation. Journal of Dairy Research 72, 287-295.

Korhonen, H. 2009. Milk-derived bioactive peptides: From science to applications. Journal of Functional Foods 1, 177-187.

Korhonen, H., Pihlanto-Leppälä, A., Rantamäki, P. and Tupasela, T. 1998. Impact of processing on bioactive proteins and peptides. Trends in Food Science and Technology 9, 307-319.

McSweeney, P. L. H. 2004. Biochemistry of cheese ripening: Introduction and overview. In: Cheese: Chemistry, Physics and Microbiology, Fox, P. F., McSweeney, P. L. H., Cogan, T. M., and Guinee, T. P. (Eds.). Academic Press, Volume 1, pp. 347-360.

Mendia, C., Ibañez, F. J., Torre, P. and Barcina, Y. 2000. Effect of pasteurization and use of a native starter culture on proteolysis in a ewes' milk cheese. Food Control 11, 195-200.

Minkiewicz, P. D., Jerzy, D., Iwaniak, A., Dziuba, M. and Darewicz, M. 2008. BIOPEP Database and other programs for processing bioactive peptide sequences. Journal of AOAC International 91, 965-980.

Nagpal, R., Behare, P., Rana, R., Kumar, A., Kumar, M., Arora, S., Morotta, F., Jain, S. and Yadav, H. 2011. Bioactive peptides derived from milk proteins and their health beneficial potentials: an update. Food and Function 2, 18-27.

Old, W. M., Meyer-Arendt, K., Aveline-Wolf, L., Pierce, K. G., Mendoza, A., Sevinsky, J. R., Resing, K. A. and Ahn, N. G. 2005. Comparison of label-free methods for quantifying human proteins by shotgun proteomics. Molecular and Cellular Proteomics 4, 1487-1502.

Pham, T. V., Piersma, S. R., Warmoes, M. and Jimenez, C. R. 2010. On the beta-binomial model for analysis of spectral count data in label-free tandem mass spectrometry-based proteomics. Bioinformatics 26, 363-369.

Raynal-Ljutovac, K., Park, Y. W., Gaucheron, F. and Bouhallab, S. 2007. Heat stability and enzymatic modifications of goat and sheep milk. Small Ruminant Research 68, 207-220.

Rizzello, C. G., Losito, I., Gobbetti, M., Carbonara, T., De Bari, M. D. and Zambonin, P. 2005. Antibacterial activities of peptides from the water-soluble extracts of Italian cheese varieties. Journal of Dairy Science 88, 2348-2360.

Singh, H. 2004. Heat stability of milk. International Journal of Dairy Technology 57, 111-119.

Tanca, A., Addis, M. F., Pagnozzi, D., Cossu-Rocca, P., Tonelli, R., Falchi, G., Eccher, A., Roggio, T., Fanciulli, G. and Uzzau, S. 2011. Proteomic analysis of formalin-fixed, paraffin-embedded lung neuroendocrine tumor samples from hospital archives. Journal of Proteomics 74, 359-370.

Vijayakumar, V., Guerrero, A. N., Davey, N., Lebrilla, C. B., Shields, D. C. and Khaldi, N. 2012. EnzymePredictor: A Tool for Predicting and Visualizing Enzymatic Cleavages of Digested Proteins. Journal of Proteome Research, 11 6056-6065.

The Grana Padano microbiota: insights from experimental caseification

Alessio Soggiu[1*], Cristian Piras[1], Stefano Levi Mortera[2,3], Milena Brasca[4], Andrea Urbani[2,3], Luigi Bonizzi[1] and Paola Roncada[1,5]

[1]Dipartimento di Scienze Veterinarie e Sanità Pubblica (DIVET), Università di Milano, Milano, Italy; alessio.soggiu@unimi.it
[2]Department of Experimental Medicine and Surgery, University of Rome 'Tor Vergata', Rome, Italy
[3]Proteomic and Metabolomic Laboratory Santa Lucia Foundation-IRCCS, Rome, Italy
[4]Istituto di Scienze delle Produzioni Alimentari (ISPA), Consiglio Nazionale delle Ricerche (CNR), Milano, Italy
[5]Istituto Sperimentale L. Spallanzani, Milano, Italy

Introduction

Bacteria are key players in cheese production because texture, flavour, nutritional properties and defects strictly depends from the dynamics of microorganisms inside the cheese during the ripening. In Grana Padano, a typical Italian cheese, is present a complex bacterial ecosystem deriving partially from the raw milk and the starter mixture added during the manufacturing step (Neviani et al. 2013). In 15-35% percent of cases during the ripening, several bacterial species belonging to the clostridia genus, in particular *Clostridium tyrobutyricum*, and deriving from endospores naturally present in raw milk causes the blowing of the cheese due to the butyric fermentation process (Klijn et al. 1995). To avoid the clostridial growth, is added, during the production, the egg lysozyme, a natural bacteriolytic enzyme very effective at low ppm concentration. Moreover, the use of lysozyme from egg is one of the potential problems in Grana Padano production as is suspected to cause potential allergic reactions. Currently are under investigation several alternatives to counteract the blowing phenomenon without the use of additives using only probiotic bacteria.

Aims

The current investigation has as its primary objective the characterization of the microbiota profile in different samples of grana padano cheese derived from an experimental caseification process. Two different variables has been taken in account: low or high clostridial spore number and lysozyme treatment or not. A metaproteomic approach was used to determine the global bacterial profile associated to each treatment to unravel the different microbioma linked to each experimental group and highlight the most important differences at the functional level.

Material and methods

Experimental cheeses made in a pilot plant were classified on the basis of the presence of lysozyme and clostridial spore number (Table 1). Thirty-hundred mg of grated Grana Padano cheese were dispersed in 1 ml of MilliQ water (Millipore, Germany) and homogenized with ultrasounds (Bandelin SONOPULS, Germany) for 2 min at 50% power and further stirred for 1 h at 40 °C. Cheese samples were centrifuged at 10,000×g for 10 min at 20 °C, to separate the aqueous phase from the pelleted caseins. The aqueous phase was precipitated with chloroform/methanol/water 4:4:1 (v/v/v) and pellet resuspended in 8 M urea. After total protein amount quantification by Bradford assay small volumes containing 50 µg of proteic material were picked and diluted to reduce concentration of urea to 2 M. 2 µl of 0.1 M DTT, 2.5 µl of 0.2 M IAA and again 0.5 µl of 0.1 M DTT were subsequently added to samples, leaving the mixture for 1 h at 37 °C, 1 h at r.t. in dark, and 20 min at 37 °C respectively. 1 µl of trypsin (0.5 µg/µl) was added; digestion was left proceeding over night at 37 °C and stopped with 2 µl of 1% TFA. 6 µl of tryptic peptides solution (3 µg) were loaded on an SCX column (2 cm, 100 µm i.d., IDEX) directly connected with a RP precolumn C18 (2 cm, 100 µm i.d., IDEX) and a C18-Acclaim PepMap column (25 cm, 75 µm i.d., 5 µm p.s., Thermo Fisher Scientific). After 3 min of preconcentration with 100% H_2O, 0,1% F.A. flow (3 µl/min) a first gradient was run to elute peptides unbounded to SCX resin (from 3 to 30% ACN in 120 min). Three salt bumps were injected to progressively elute charged peptides (18 µl, 10, 100, and 500 mM respectively) with the same gradient as previous. The analytical column was connected with the nano-spray source of a Bruker amAzon ETD Ion Trap working in Auto MSn mode (DDA) recording 10 MSMS spectra for each survey scan. Raw data were processed with Compass Data Analysis 1.3 (Bruker Daltonics) while protein identifications and 2D fractions data combining was performed with Compass Proteinscape 2.1 (Bruker Daltonics) against the NCBInr database restricted to bacteria. The identified peptide list was converted to a fasta file using a pyton script. Each xml dataset was compared to a reference bacterial database using Blast 2.2.9 and the results imported directly to MEGAN 5.5.3(Huson and Weber 2013) for the taxonomical and the functional classification of the reads.

Results and discussion

A total of eight samples were processed for the metaproteomics analysis (Table 1).

Table 1. Experimental cheese sample processed for metaproteomics analysis.

Sample	Code	Clostridial spore count	Lysozyme
1	F27a	low	–
2	F27b	high	–
3	F28a	low	–
4	F28b	high	–
5	F29a1	low	+
6	F29b1	high	+
7	F30a1	low	+
8	F30b1	high	+

After the metaproteomics analysis about eight hundred reads were obtained for each sample with slight differences in terms of the total number of reads. Bacterial peptides were assigned to bacterial phyla using the bioinformatic pipeline described in the methods paragraph. In all samples is possible to define a shared core microbiome represented by (in order of relative abundance): *Lactobacillus*, *Streptococcus*, Enterobacteriacee, Actinomycetales, Bacteroidetes, Clostridiales, Rhizobiales, Deltaproteobacteria, Halella, Rhodospirillum, Burkholderiales and Bacillales. As expected, the first three more abundant genera reflect the bacterial profile of the whey starter mixture used initially in the cheese-making process; clostridia and the other less abundant genera most likely have an environmental origin (raw milk, manufacturing facilities). Using the SEED functional subsystem, we were able to highlight, in absence of lysozyme, key differences that depend only by the number of clostridial spore present. In particular carbohydrates, stress response, RNA, aminoacid and nitrogen metabolism are up-regulated in the high spore samples and protein, cell wall and DNA metabolism are up-regulated in low spore samples. In the high spore samples, the lysozyme treatment switch-off some pathways previously up-regulated like the RNA metabolism and in parallel the same pathway and the motility and chemotaxis profile is switched-on in the low spore samples. At the moment other analyses are in progress for the investigation of those interesting functional changes at the protein and microbial specie level to elucidate and define new production strategies in the control of the blowing phenomena in the Grana Padano cheese without the use of additives

Acknowledgements

Work supported by the FILIGRANA project, financed from MiPAAF D.M 25741/7303/11 – 01/12/2011.

References

Huson, D. H., and N. Weber. 2013. Microbial community analysis using MEGAN. *Methods Enzymol* 531: 465-85. http://dx.doi.org/10.1016/B978-0-12-407863-5.00021-6.

Klijn, N., F. F. Nieuwenhof, J. D. Hoolwerf, C. B. van der Waals, and A. H. Weerkamp. 1995. Identification of Clostridium tyrobutyricum as the causative agent of late blowing in cheese by species-specific PCR amplification. *Appl Environ Microbiol* 61 (8): 2919-24.

Neviani, E., B. Bottari, C. Lazzi, and M. Gatti. 2013. New developments in the study of the microbiota of raw-milk, long-ripened cheeses by molecular methods: the case of Grana Padano and Parmigiano Reggiano. *Front Microbiol* 4: 36. http://dx.doi.org/10.3389/fmicb.2013.00036.

Supplemented diets for allergenic modulation in European seabass

Denise Schrama[1*], Jorge Dias[2] and Pedro M. Rodrigues[1]

[1]CCMAR, Universidade do Algarve, Edifício7, Campus de Gambelas, 8005-139 Faro, Portugal; dschrama@ualg.pt

[2]Sparos Lda, Area Empresarial de Marim, Lote C, 8700-221 Olhão, Portugal

Introduction

Food allergy is becoming a common health problem worldwide with more than 17 million European (FARE, 2014) having some kind of food allergy. Proteins are in most cases the molecules responsible for this type condition in food and may cause severe allergenic reactions due to abnormal responses of the immune system (Taylor, 2000), classified as Immunoglobulin E (IgE) or non IgE mediated reaction (Waserman and Watson, 2011).

Fish allergy affects approximately 0.4% of the population and the best known allergen is parvalbumin, being highly abundant and stable (Swoboda *et al.*, 2002). Parvalbumin is a calcium-binding protein with a small molecular weight around 12 kDa, and depending on its evolution has a pI of 4.5 (β group) or 5.0 (α group) (Swoboda *et al.*, 2002). This protein shows highly conserved regions, resulting in cross-reactions of parvalbumin between different fish species. Other allergens were also identified as triggers for an allergic reaction, like collagen (identified in various fish species; Hamada *et al.*, 2001), vitellogenin (raw caviar; Perez-Gordo *et al.*, 2008), aldehyde phosphatase dehydrogenase (cod; Das Dores *et al.*, 2002) and gelatin.

Creatine is a molecule which supplies the muscle and other cells in the body with energy and is widely used by athletes as a supplement to gain muscle mass. The European Food Safety Authority considered an intake of 3 g per day as safe. Previous studies performed in rat to see the effect on the parvalbumin content when adding creatine to diets, show a 75% decrease in parvalbumin content (Gallo *et al.*, 2008).

In this study we tried to address for the first time the effect of creatine in the muscle proteome of European seabasss with particular interest in the expression of fish major allergen parvalbumin, using food diets with different concentrations of creatine in its composition.

Methodology

Two diets supplemented with different creatine concentrations (0.35 and 2%) were studied against a control diet in European seabass (*Dicentrarchus labrax*) for 57 days at the experimental station (Ramalhete) of the University of Algarve. Thirty fish (mean initial body weight of 73±0.8 g) were kept in 500 l tanks with natural ranges of temperature (19.7±2.4 °C), photoperiod, artificial aeration (dissolved oxygen above 5 mg/l), salinity and a rearing density of 4.3 kg/m^3. Fish were fed till apparent satiety. Muscle samples were taken at the end of the trial. Proteins were extracted from 50 mg of muscle sample of 5 fish per condition in DIGE buffer and quantified by the Bradford method. Fifty µg of protein were labeled with either CyDye3 or CyDye5 with a pool of samples being labeled with CyDye2 as internal standard. After a half hour incubation the reaction was stopped by adding Lysine (10 mM). Muscle proteins were focused on 24 cm Immobiline DryStrips (GE Healthcare, Sweden) with pH 3-10 NL until a total of 60,000 Vhr. A reduction and alkylation step was done with DTT and iodoacetamide, respectively before separation according to molecular weight on 12.5% polyacrylamide gels. Gels were scanned on a 9400 Typhoon scanner (GE Healthcare, Sweden) and analyzed with SameSpots (Totallab, United Kingdom). Significantly different spots ($P<0.05$ by ANOVA) and spots known to be in the parvalbumin region were excised manually and sequenced by MALDI-TOF. Creatine and parvalbumin contents in muscle were analyzed using commercial kits (MAK079, Sigma Aldrich and R6010-2E, Bio-Check UK, respectively).

Results and discussion

A gel obtained from this analysis is represented in Figure 1 with some of the sequenced spots highlighted.

We were able to identify the fish allergen in the acidic, small molecular weight region. In this trial we were interested in identifying this protein, so although it wasn't significantly different expressed with creatine supplementation we were able of identify parvalbumin in 2D gels and by western blot (data not shown), giving us more information for further analysis.

Spot 405 was identified as a fast white muscle troponin which is structurally related to parvalbumin and some isoforms are also able to bind to calcium. The muscle type creatine kinase CKM1 was

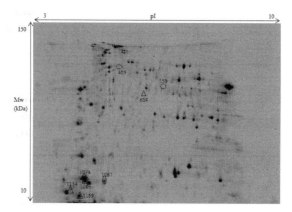

Figure 1. 2D-DIGE gel from muscle protein of European seabass. Triangle spots are up-regulated in control fish, circled spots are up-regulated in fish fed the creatine 0.35% diet and squared spots are up-regulated in fish fed the creatine 2% diet.

identified in spot 550, which catalyzes the conversion of creatine to phosphocreatine consuming ATP (adenosine triphosphate) and generates ADP (adenosine diphosphate).

To understand if the amount of creatine and parvalbumin in muscle samples altered due to diet supplementation, analysis with kits were performed. The results are shown in Figure 2A and 2B, respectively. No significant differences ($P>0.05$) were observed between conditions. These results seem to point that for these levels of creatine used no particular effect is observed, either in parvalbumin expression or its accumulation as most probably all the creatine is being excreted to blood as creatinine. Further analysis need to be carried out to further understand the creatine effect on fish muscle.

Future work

Different/higher creatine supplementation concentrations will be used in a new trial with Gilthead seabream to see the effect on the parvalbumin content in muscle. Various zootechnical and biochemical data will also be analyzed to address growth performances and understand creatine role on fish metabolism.

Acknowledgements

This work is part of project 31-03-05-FEP-0060 – Aquavalor, co-financed by PROMAR: Projetos Pilotos e a Transformação de embarcações de Pesca.

Denise Schrama acknowledges scholarship on project Aquavalor, Refª 31-03-05-FEP-0060.

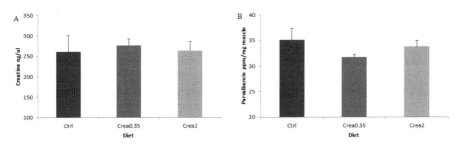

Figure 2. (A) Creatine levels (ng/µl) in muscle of European seabass. (B) Parvalbumin levels (ppm/mg) in muscle of European seabass (Control – Ctrl, Creatine 0.35% – Crea0.35, Creatine 2% – Crea2). Means (n=5) with standard deviation represented as error bars.

References

Das Dores, S., Chopin, C., Romano, A., Galland-Irmouli, A.V., Quaratino, D., Pascual, C., Fleurence, J. and Gueant, J.L., 2002. IgE-binding and cross-reactivity of a new 41 kDa allergen of codfish. Allergy 57: 84-87.

FARE, 2014. Tools & resources, facts and statistics. Food Allergy Research & Education McLean, Virginia, United States.

Gallo, M., MacLean, I., Tyreman, N., Martins, K.J.B., Syrotuik, D., Gordon, T. and Putman, C.T., 2008. Adaptive responses to creatine loading and exercise in fast-twitch rat skeletal muscle. American Journal of Physiology-Regulatory Integrative and Comparative Physiology 294: R1319-R1328.

Hamada, Y., Nagashima, Y. and Shiomi, K., 2001. Identification of collagen as a new fish allergen. Bioscience Biotechnology and Biochemistry 65: 285-291.

Perez-Gordo, M., Sanchez-Garcia, S., Cases, B., Pastor, C., Vivanco, F. and Cuesta-Herranz, J., 2008. Identification of vitellogenin as an allergen in Beluga caviar allergy. Allergy 63: 479-480.

Swoboda, I., Bugajska-Schretter, A., Verdino, P., Keller, W., Sperr, W.R., Valent, P., Valenta, R. and Spitzauer, S., 2002. Recombinant carp parvalbumin, the major cross-reactive fish allergen: A tool for diagnosis and therapy of fish allergy. Journal of Immunology 168: 4576-4584.

Taylor, S.L., 2000. Emerging problems with food allergens. Food and Agriculture Organization, Rome, Italy.

Waserman, S. and Watson, W., 2011. Food allergy. Allergy Asthma Clin Immunol 7 Suppl 1: S7.

Protein changes in muscles with varying meat tenderness

Eva Veiseth-Kent[*], Kristin Hollung, Vibeke Høst and Rune Rødbotten
Nofima – the Norwegian Institute of Food, Fisheries and Aquaculture Research, P.O. Box 210, 1431 Ås, Norway; eva.veiseth-kent@nofima.no

Introduction

Tenderness is a critical factor determining the consumer's acceptance of meat, and unfortunately considerable variation in tenderness is found between different cuts and muscles. In addition, studies have shown that feed can have an effect on meat tenderness (May *et al.*, 1992). One of the main determinants of meat tenderness is calpain-mediated proteolysis of key myofibrillar and associated proteins during *post-mortem* cooler storage of meat (Huff-Lonergan *et al.*, 2010). These proteins maintain the structural integrity of the myofibrils, and once degraded the rigid structure of the myofibrils is weakened leading to muscle fibre breakage and more tender meat (Veiseth-Kent *et al.*, 2010). Moreover, a possible role of matrix metalloproteases (MMPs) in the breakdown of connective tissue in meat has been suggested (Purslow *et al.*, 2012). The expression of MMPs in muscle fibres are also found to depend on the muscle fibre type (Cha and Purslow, 2010), and could therefore potentially play a role in explaining some of the variation in tenderness between different bovine muscles. The objective was therefore to assess meat tenderness, general protein changes by 2D-DIGE proteome analysis, and *post-mortem* proteolysis (measured as calpain-specific degradation of Troponin-T (TnT) and activity of the pro and active form of MMP-2) in four different muscles; *Longissimus dorsi* (LD), *Rectus femoris* (RF), *Infraspinatus* (IS), and *Vastus lateralis* (VL), from Norwegian Red heifers subjected to fertilized conventional pasture or natural land with no fertilizer.

Materials and methods

Animals and sampling

Twelve Norwegian Red heifers were either subjected to fertilized conventional pasture (Group 1) or natural land with no fertilizer (Group 2). All animals were born in December and had two summers on pasture. The animals were slaughtered at a commercial abattoir, and samples were collected from four different muscles (LD, RF, IS, VL) following 2 and 14 days of cooler storage. At both time points, samples were snap frozen in liquid nitrogen and stored at -80 °C for determination of TnT-degradation, MMP-2 activity and 2D-DIGE analysis, while WBSF was measured on fresh samples at 14 d *post-mortem*. In addition, samples from the LD muscle were also snap frozen 1 h after slaughter for 2D-DIGE.

Protein analyses

For the proteome analysis, water-soluble proteins from 1 h and 14 d *post-mortem*, and the insoluble protein fractions from 14 d *post-mortem* were analysed by 2D-DIGE (GE Healthcare, USA). Gel images were analysed using the Progenesis SameSpots software (version 4.5; Nonlinear Dynamics, UK), and statistical analysis of the spot volumes were performed using the q-value calculation in SameSpots and with Partial Least Squares Regression (PLSR) with Jack-knife. Significantly altered protein spots were identified by MALDI-TOF/TOF mass spectrometry. Degradation of TnT was measured as the occurrence of a calpain-specific 30-kDa fragment analysed by Western blot, while the activity of MMP-2 was measured by gelatin-zymography.

Meat tenderness

For Warner-Bratzler shear force (WBSF) measurements, slices of all the muscles (3.5 cm thick) were vacuum packed, heated in a water bath at 70 °C for 50 min and chilled in iced water for 45 min. Small samples were then cut from the cooked meat with slices parallel to the fibre direction. This produced 10 samples (2×1×1 cm) which were sheared using a WBSF triangular version device mounted in an Instron Materials Testing Machine. The average results from the 10 samples were used for each muscle in the statistical analysis.

Statistical analysis

Analysis of variance for WBSF, TnT degradation and MMP-2 activity was performed using MINITAB's general linear model (Minitab, version 16.1.1; Minitab Ltd., UK). The models for TnT and MMP-2 included the following factors: feeding group, muscle, ageing period, and all two-factor interactions. The model for WBSF included the following factors: feeding group, muscle, and their interaction. The significant level was set to $P<0.05$, and when the effect of a factor was significant, means were separated using the Tukey method.

Results and discussion

Results from the WBSF measurements showed a significant interaction between feeding group and muscle (Figure 1), and the IS muscle had lower WBSF values (i.e. was more tender) compared to all other muscles for group 1, and lower values compared to the VL from group 2. There were no differences between the feeding groups in WBSF within the individual muscles.

From the 2D-DIGE analysis of the water-soluble proteins from LD muscles at 1 h and 14 d *post-mortem*, no significant changes were found between the two animal groups however, 52 protein spots were significantly changed in abundance between the two time points at 5% significance level (Figure 2). Of these protein spots, 16 were successfully identified by MALDI-TOF/TOF mass spectrometry (Table 1). In the samples collected at 1 h *post-mortem* we observed an elevated abundance of small heat shock proteins, enzymes involved in glycolysis and energy metabolism

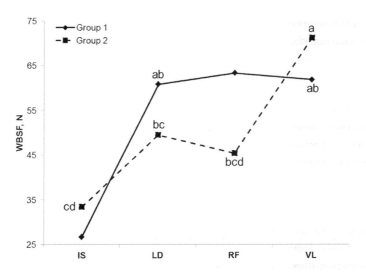

Figure 1. Warner-Bratzler shear force in four different muscles from two feeding groups measured at 14 d post-mortem. Means with different letters are significantly different (P<0.05).

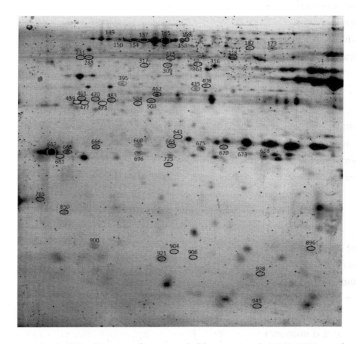

Figure 2. Representative 2-DE gel image of water-soluble proteins (pI 5-8, 12% acrylamide). Proteins altered in abundance between 1 h and 14 d post-mortem are marked.

Table 1. Proteins showing significant alternation in abundance from 1h to 14d post-mortem *in the* Longissimus dorsi *muscle.*

Spot no.	Protein name (source: *Bos taurus*)	NCBI Acc. no.	Matched pep. / %coverage	Ratio 1 h:14 d
149	Heat shock protein 70	Gi|163310909	12 / 37	0.1
150	Stress-70 protein	Gi|77735995	13 / 23	0.6
154	Serum albumin	Gi|367460260	13 / 22	0.7
157	Serum albumin	Gi|367460260	15 / 23	0.7
165	Serum albumin	Gi|367460260	9 / 16	0.7
173	Phosphoglucomutase-1	Gi|116004023	23 / 43	0.6
316	Alpha-enolase	Gi|528979842	17 / 39	1.9
395	Ankyrin-repeat domain protein	Gi|528926411	11 / 30	5.7
435	Isocitrat dehydrogenase	Gi|296475389	10 / 33	2.5
660	Heat shock protein beta-1	Gi|71037405	9 / 64	4.6
662	Myosin light chain 1/3 skeletal muscle isoform	Gi|1181841	10 / 61	9.3
673	Heat shock protein beta-1	Gi|71037405	11 / 60	12.0
675	Heat shock protein beta-1	Gi|71037405	10 / 62	15.0
688	Adenylate kinase isoenzyme 1	Gi|61888850	18 / 70	0.2
696	Myosin light chain 6B	Gi|115496556	12 / 55	7.7
900	Phosphohistidine phosphatase	Gi|115497372	8 / 44	2.8

and myosin light chain as compared to the 14-day samples. At 14 d *post-mortem* heat shock protein 70, serum albumins, phosphoglucomutase-1 and adenylate kinase had elevated abundances. These changes agree with several other proteome studies of *post-mortem* changes in muscle, showing increased energy metabolism and oxidative stress conditions during the very early *post-mortem* stages (Paredi *et al.*, 2013). Regarding the insoluble protein fraction, no differences where observed between the two feeding groups after 14 d storage.

Occurrence of the calpain-specific 30-kDa TnT fragment was similar at 2 d in the four muscles, but LD had the highest level of proteolysis at 14 d *post-mortem* (Figure 3A). MMPs exist in two different forms *in vivo*; an inactive pro-form (pro-MMP-2) and an active form (MMP-2). For pro-MMP-2, LD had lower activity at 2 d *post-mortem* compared to RF and VL, while no differences were seen after 14 days of ageing (Figure 3B). For the active MMP-2, the activity was higher in animals subjected to fertilized conventional pasture compared to natural land, and reduced at 14 d compared to 2 d *post-mortem*. Interestingly, the increased MMP-2 activity for animals on fertilized conventional pasture did however not lead to any detectable changes in the insoluble protein fraction analysed by 2D-DIGE. This could however be related to either the sensitivity or the selectivity of the proteomics approach used in this study.

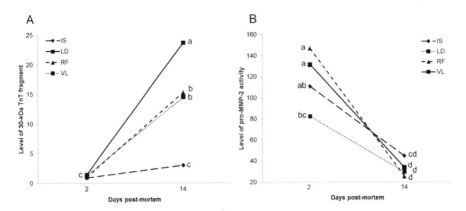

Figure 3. Occurrence of the 30-kDa Troponin-T fragment in four different muscles at 2 d and 14 d post-mortem (A), and the pro-MMP-2 activity in four different muscles at 2 d and 14 d post-mortem (B). Means with different letters are significantly different (P<0.05).

Conclusion

No differences were observed in the protein abundance between the two feeding groups, and no increased proteolysis was detected in the insoluble protein fraction after 14 d storage. The *post-mortem* proteolytic activity varied between the LD, IS, VL and RF muscles, however the variation cannot explain all differences in meat tenderness observed between the muscles. Information on *post-mortem* tenderization potential for different beef muscles can provide useful knowledge regarding optimal handling and storage procedures for different beef muscles.

Acknowledgements

Financial support for this study was provided by grants from the Norwegian Research Council (199406/199) and the Fund for Research Levy on Agricultural Products in Norway.

References

Cha, M.C. and Purslow, P.P., 2010. Matrix metalloproteinases are less essential for the in situ gelatinolytic activity in heart muscle than in skeletal muscle. Comparative Biochemistry and Physiology Part A: Molecular & Integrative Physiology 156:518-522.

Huff-Lonergan, E., Zhang, W. and Lonergan, S.M., 2010. Biochemistry of *postmortem* muscle – Lessons on mechanisms of meat tenderization. Meat Science 86:184-195.

May, S.G., Dolezal, H.G.,Gill, D.R., Ray, F.K. and Buchanan, D.S., 1992. Effects of days fed, carcass grade traits, and subcutaneous fat removal on postmortem muscle characteristics and beef palatability. Journal of Animal Science 70: 444-453.

Paredi, G., Sentandreu, M., Mozzarelli, A., Fadda, S., Hollung, K. and de Almeida, A.M., 2013. Muscle and meat: New horizons and applications for proteomics on a farm to fork perspective. Journal of Proteomics 88:58-82.

Purslow, P.P., Archile-Contreras, A.C. and Cha, M.C., 2012. Manipulating meat tenderness by increasing the turnover of intramuscular connective tissue. Journal of Animal Science 90: 950-959.

Veiseth-Kent, E., Hollung, K., Ofstad, R., Aass, L. and Hildrum, K.I., 2010. Relationship between muscle microstructure, the calpain system, and shear force in bovine *longissimus dorsi* muscle. Journal of Animal Science 88:3445-3451.

Intact cell MALDI-TOF MS: a male chicken fertility phenotypic tool

Laura Soler[1,2,3,4]*, Aurore Thèlie[1,2,3,4], Valérie Labas[1,2,3,5], Ana-Paula Teixeira-Gomes[5,6,7], Isabelle Grasseau[1,2,3,4], Grégoire Harichaux[1,2,3,4,5] and Elisabeth Blesbois[1,2,3,4]

[1]*INRA, UMR85 Physiologie de la Reproduction et des Comportements, 37380 Nouzilly, France; lsolervasco@tours.inra.fr*
[2]*CNRS, UMR7247, 37380 Nouzilly, France*
[3]*Université François Rabelais de Tours, 37000 Tours, France*
[4]*IFCE, Institut Français du Cheval et de l'Equitation, 37380 Nouzilly, France*
[5]*INRA, Plate-forme d'Analyse Intégrative des Biomolécules, Laboratoire de Spectrométrie de Masse, 37380 Nouzilly, France*
[6]*INRA, UMR 1282 Infectiologie et Santé Publique, 37380 Nouzilly, France*
[7]*Université François Rabelais de Tours, UMR1282 Infectiologie et Santé Publique, 37000 Tours, France*

Objectives

Animal breeding often faces the paradox of choosing between growth and reproduction, which are inversely related. Adequate nutritional and management schemes together with genetic control has helped advancing in this field, but molecular tools still need to be developed for fast, reliable screening of good breeders. In chicken, the evaluation of the reproductive ability on an individual can be accurately done by measuring the egg fertilization rate and tracking the embryonic development through a standardized process. However, although the latter is more precise than *in vitro* sperm quality tests (which are faster and simpler), its routine use is cumbersome and expensive, and therefore not suitable for screening purposes. Often, results of both tests are not correlated, as good fertility rates can be obtained when inseminating hens with sperm of low *in vitro* quality, and vice-versa. Recently, we developed a new tool based on intact cell MALDI-TOF mass spectrometry (ICM-MS), which was tested in a pilot study and proven promising for the evaluation of fertility in male chicken sperm cells (Labas *et al.*, 2014). Here, this method was technically optimized and later applied to a larger number of animals to evaluate the proteomic differences between fertile and sub-fertile spermatozoa from a broiler chicken breed.

Material and methods

Twenty roosters were classified depending on their reproductive quality based on *in vitro* and *in vivo* fertility tests. Fresh cells (three ejaculates per male) were washed twice in Tris-Sucrose buffer and resuspended at 10^9 cells/ml. One microliter of cell suspension (10^6 cells) was mixed with 2.5 µl of matrix (20 mg/ml Sinapinic acid, 2% Trifluoroacetic acid, 50% acetonitrile in water) and 12 replicates were spotted onto a MTP Ground Steel 384 MALDI plate (Bruker Daltonics, Germany). Spectra were acquired three consecutive times using a Bruker UltrafleXtreme MALDI-TOF-TOF instrument

(Bruker Daltonics, Germany) in the mass range of 2-20 kD. External calibration was followed using a mixture of peptides and proteins containing Glu1-fibrinopeptide B, ACTH (fragments 18-39), insulin and ubiquitin, all at 1 pmol/µl, 2 pmol/µl cytochrome C, 4 pmol/µl myoglobin and 8 pmol/µl trypsinogen. Using flexAnalysis software (Bruker Daltonics, Germany), each spectrum was converted and saved in a text file. Spectral processing and analysis was performed with Progenesis MALDI v1.2 software (Nonlinear Dynamics, UK). To characterize m/z peak differences between samples, the mean normalized peak height intensity values were subjected to one way analysis of variance (ANOVA), and principal component analyses (PCA) were processed using peaks with $P<0.01$ and more than two-fold change. To ensure the statistical relevance of the results generated, we checked that 100% of the data had a power >0.8.

Targeted proteomic analysis using top-down MS approach was performed to identify significant different m/z peaks ($P<0.01$) between fertile and sub-fertile males. In brief, peptides/proteins were extracted from sperm cells by sonication in 6 M Urea 50 mM Tris-HCl pH 8.8 and subjected to chromatographic separation on an UltiMate 3000 RSLC system controlled by Chromeleon version 6.80 SR13 software (Thermo Scientific Dionex, USA). Two different chromatographic approaches were employed to fractionate proteins and peptides. First, proteins were separated by reversed phase HPLC using a Waters XBridge BEH C18 column (Waters, USA). A second approach was based in the separation by molecular weight using a Superdex 75 10/300 GL gel filtration column (GE Healthcare Life Sciences, USA) in 100 mM ammonium bicarbonate buffer. Fractions obtained from each chromatographic procedure were dried, resuspended in formic acid 1% and analyzed by MALDI-TOF to identify masses of interest. All fractions were then analyzed by on-line nanoflow liquid chromatography tandem mass spectrometry (nanoLC-MS/MS). All experiments were performed on a dual linear ion trap Fourier Transform Mass Spectrometer (FT-MS) LTQ Orbitrap Velos (Thermo Fisher Scientific, Germany) coupled to an Ultimate® 3000 RSLC Ultra High Pressure Liquid Chromatographer (Dionex, the Netherlands) controlled by Chromeleon Software (version 6.8 SR11; Dionex, the Netherlands). The separation was conducted using a Dionex column (Monolithic PS-DVB PepSwift) in an acetonitrile gradient. Data were acquired using Xcalibur software (version 2.1; Thermo Fisher Scientific, USA). Proteo/peptidoform identification and structural characterization were performed using ProSight PC software v 3.0 SP1 (Thermo Fisher, USA). Automated searches were performed using the 'Biomarker' search option against a database made via shotgun annotation from the Swiss-Prot Gallus gallus release from the UniProtKnowledgebase release 2012_06.

Results and discussion

Fertile and sub-fertile males were successfully discriminated based on their MS profiles, thus confirming the suitability of Intact Cells MALDI-TOF MS to evaluate fertility. As expected, the results obtained from sperm quality *in vitro* tests and from *in vivo* fertility tests were not well correlated in some cases. These differences were reflected in the observed proteomic profile, allowing us to interpret which molecular events are effectively related with subfertility in chicken. Remarkably, most of the molecular signatures related with subfertility appeared in the low mass range (approx. 5-7 kD) of the spectra of subfertile cells (Figure 1). The top-down protein identification approach used here combined with two different chromatographic pre-fractionation protocols allowed us to maximize the number of reliable protein identifications compared with past attempts. We were therefore able to better characterize the chicken sperm cells proteomic profile, and to identify different peptido/proteoforms (endogenous species) related to impaired fertility.

In conclusion, ICM-MS can be considered as an innovative 'phenomic molecular tool' that may help discriminating avian males on their reproductive capacity. Furthermore, ICM-MS combined to chromatographic fractionation and top-down protein identification allowed characterization of several new molecular species related to fertility.

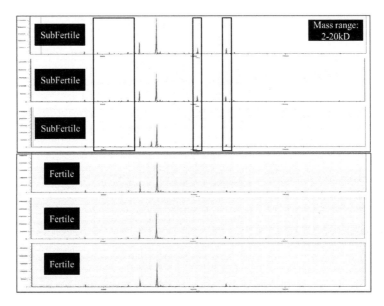

Figure 1. Representative intact spermatozoa MALDI-TOF MS profiles from 3 subfertile and 3 fertile males. The regions showing visually evident differences are boxed. Mass range was 2-20 kD.

Acknowledgements

This research was sponsored by the project CRB-Anim. Laura Soler has received the support of the European Union, in the framework of the Marie-Curie FP7 COFUND People Programme, through the award of an AgreenSkills' fellowship (under grant agreement n° 267196). Funding was also received from the SMHART project by the European Regional Development Fund (ERDF), the Conseil Régional du Centre, the French National Institute for Agricultural Research (INRA) and the French National Institute of Health and Medical Research (Inserm).

References

Labas, V., Grasseau, I., Cahier, K., Gargaros, A., Harichaux, G., Teixeira-Gomes, A.P., Alves, S., Bourin, M., Gerard, N. and Blesbois, E., 2014. Qualitative and quantitative peptidomic and proteomic approaches to phenotyping chicken semen. J Proteomics.

Identification of novel protein interactions related to spermatid elongation

Mari Lehti[1], Noora Kotaja[2] and Anu Sironen[1]*
[1]Agrifood Research Finland, Biotechnology and Food Research, Animal Genomics, 31600 Jokioinen, Finland; anu.sironen@mtt.fi
[2]Department of Physiology, Institute of Biomedicine, University of Turku, 20520 Turku, Finland

Background

An infertility defect was identified in the Finnish Large White pig population in early 1990's causing male infertility due to short and immotile sperm. Genome mapping and candidate gene sequencing identified the causal mutation for the defect as a L1 insertion within Spef2 gene (Sironen *et al.*, 2007). The role of this gene was studied in the mouse and identified an interaction with IFT20 through yeast-two-hybrid-analysis (Sironen *et al.*, 2010). This interaction indicates an involvement of SPEF2 protein in the protein transport mechanisms during spermatogenesis. Since not much is known about the intra flagellar transport (IFT) during spermatid elongation, we developed a male germ cell specific conditional knock-out (KO) mouse model for a known IFT motor protein, Kif3A and for Spef2. The phenotypes of these mouse models are very similar with lack of the sperm tail and malformed manchette. Thus, these phenotypes indicate a possible role for these proteins in the manchette and sperm tail formation/function.

Novel interaction candidates for Kif3A

Co-immunoprecipitation (Co-IP) from adult WT testis lysates using KIF3A and negative control antibody (Rabbit IgG), followed by mass spectrometric analysis, was performed to identify possible interacting partners for KIF3A in the testis (Lehti *et al.*, 2013). Co-IP was done in triplicate to eliminate false positive results. The previously reported interactions between KIF3A and other kinesin II subunits, KIF3B (Yamazaki *et al.*, 1995), and KAP3 (Yamazaki *et al.*, 1996) were confirmed by our data. All three proteins were detected with high coverage of peptide sequences (29-66%) in the mass spectrometric analysis.

Furthermore, eight other interacting candidates for KIF3A were identified: MNS1 (Meiosis-specific nuclear structural protein1, Q61884); SMRP1 (Spermatid-specific manchette-related protein 1, Q2MH31); KBP (KIF1-binding protein, Q6ZPU9); Enkur (Enkurin, Q6SP97); ODF3 (Outer dense fiber protein 3, Q920N1); CCDC105 (Coiled-coil domain-containing protein 105, Q9D4K7); CCDC11 (Coiled-coil domain-containing 11, Q9D439) and FAM166A (Protein FAM166A, Q9D4K5). These interactions were identified in all three KIF3A Co-IPs while Rabbit IgG control Co-IPs did not result in any protein matches.

MNS1 localization in KIF3A KO mice

MNS1 has been shown to play an important role in sperm tail development. MNS1 KO mice have short sperm tails, the fibrous sheath is missing, and the axonema and ODFs are completely disorganized (Zhou *et al.*, 2012). This prompted us to further investigate the localization of MNS1 and the possible interaction with KIF3A. For localization of MNS1 during spermatogenesis we used two different antibodies: sc-138435 (Santa Cruz), and UP-2284 (kindly provided by PhD P. Jeremy Wang). The staining pattern was studied in paraffin embedded sections, drying down and sperm slides (Lehti *et al.*, 2013). MNS1 co-localized with KIF3A in the manchette and in the principal piece of the elongating spermatid tail. In addition, MNS1 was present in the acrosomal region and perinuclear ring. Acrosomal staining was observed in all steps of round spermatids with the signal concentrated in the acrosomal granule and the marginal areas of the acrosome. This staining was not affected in the KIF3A KO spermatids.

In the mature spermatozoa isolated from the cauda epididymis, MNS1 was found in the principal piece of the sperm tail indicating its dynamic translocation from the acrosomal region *via* the manchette to the sperm tail. Despite the localization in the acrosome and manchette, the most critical function of MNS1 seems to be associated with flagellar assembly since this is the only process that is affected in MNS1 KO spermatids (Zhou *et al.*, 2012). Interestingly, MNS1 seems to be retained in the abnormally formed manchette of the KIF3A KO elongating spermatids as well as in the acrosomal region. This suggests KIF3A functional impairment causes defects in the dynamics of MNS1 during the spermatid elongation process (Lehti *et al.*, 2013).

KBP localization during spermatogenesis

For KBP staining we used Anti-KIAA1279 antibody (Abcam). KBP was first detected sparsely in the chromatoid body (CB) of step 8 round spermatids in paraffin embebbed testis sections. Intense staining was detected during step 9 and faded in step 11-12 elongating spermatids. In drying down preparations KBP is detected until step 12-15 elongating spermatids. To confirm the KBP localization in the CB we performed double staining using TSKS as a CB marker. TSKS (testis-specific serine/threonine kinase substrate) has been shown to localize specifically in the late CB ring and satellite structure (Shang *et al.*, 2010). We were able to colocalize KBP staining with TSKS confirming its specific appearance in the late CB. In the Kif3A KO mice the CB structure appeared scattered compared to the WT mice. Co-localization of KBP and Kif3A was detected in the manchette. Thus, the Kif3A KO may affect the translocation of KBP through the manchette, thus causing the identified disintegration of the CB.

Novel Spef2 interaction candidates

Co-IP with adult WT and Spef2 KO testis lysates using SPEF2 antibody and a negative control antibody (rabbit IgG), followed by mass spectrometric analysis, was performed to identify possible interacting partners for SPEF2 in the testis. Peptides for Clathrin heavy chain 1 (CLTC), Plakophilin-1

(PKP1), Pyruvate dehydrogenase E1 component subunit beta (PDHB), Polyubiquitin-B (UBB) and 14-3-3 protein zeta/delta (YWHAZ) were identified only in the WT testis sample. In addition, high amount of peptides for Cytoplasmic dynein 1 heavy chain 1 (DYNC1H1) was identified in the WT samples and lower number in the KO. No peptides for Cytoplasmic dynein 1 heavy chain 1 were identified in the negative control. We analyzed the mRNA expression patterns of Spef2 and novel interacting partners in juvenile testis at postnatal days (PND) 7, 14, 17, 21 and 28 by using RNAseq data and the Cufflinks pipeline (Trapnell *et al.*, 2012). Each time point corresponds to the appearance of different cell populations during the first wave of spermatogenesis. Gene expression analysis showed very low expression for PKP1 during the first wave of spermatogenesis, which may indicate a false positive result. CLTC and PCHD showed highest expression at PND17, which may indicate a role in spermatid elongation. UBB showed high and increased expression (FPKM>250) throughout the sperm development. YWHAZ had peak expression at PND14, which does not correlate with SPEF2 expression. DYNC1H1 expression was highest at PND21. Although the interaction between SPEF2 and DYNC1H1 was identified in the Spef2 KO testis as well, this is a probable interaction candidate for SPEF2 due to the identified motility defect of the sperm tail and tracheal cilia. The identified interaction in the KO mice may be due to different isoforms of SPFE2. However, these interaction candidates require confirmation and co-localization studies during spermatogenesis.

Conclusions

Our previous and ongoing studies provide light to events in correct spermiogenesis and spermatid elongation process. Male infertility is an increasing problem in farm animals and investigation of defects leading to impaired spermatogenesis and the functional role of protein networks behind these defects using the mouse as a model provides crucial insights into reproduction related factors. We have identified several novel candidate proteins for male fertility using protein interaction studies.

References

Lehti, M. S., N. Kotaja, and A. Sironen. 2013. KIF3A is essential for sperm tail formation and manchette function. Mol. Cell. Endocrinol. 377(1-2): 44-55.

Shang, P., W. M. Baarends, J. Hoogerbrugge, M. P. Ooms, W. A. van Cappellen, A. A. de Jong *et al.* 2010. Functional transformation of the chromatoid body in mouse spermatids requires testis-specific serine/ threonine kinases. J. Cell. Sci. 123(Pt 3): 331-339.

Sironen, A., J. Hansen, B. Thomsen, M. Andersson, J. Vilkki, J. Toppari *et al.* 2010. Expression of SPEF2 during mouse spermatogenesis and identification of IFT20 as an interacting protein. Biol. Reprod. 82(3): 580-590.

Sironen, A., J. Vilkki, C. Bendixen, and B. Thomsen. 2007. Infertile finnish yorkshire boars carry a full-length LINE-1 retrotransposon within the KPL2 gene. Mol. Genet. Genomics. 278(4): 385-391.

Trapnell, C., A. Roberts, L. Goff, G. Pertea, D. Kim, D. R. Kelley *et al.* 2012. Differential gene and transcript expression analysis of RNA-seq experiments with TopHat and cufflinks. Nat. Protoc. 7(3): 562-578.

Yamazaki, H., T. Nakata, Y. Okada, and N. Hirokawa. 1995. KIF3A/B: A heterodimeric kinesin superfamily protein that works as a microtubule plus end-directed motor for membrane organelle transport. J. Cell Biol. 130(6): 1387-1399.

Yamazaki, H., T. Nakata, Y. Okada, and N. Hirokawa. 1996. Cloning and characterization of KAP3: A novel kinesin superfamily-associated protein of KIF3A/3B. Proc. Natl. Acad. Sci. U. S. A. 93(16): 8443-8448.

Zhou, J., F. Yang, N. A. Leu, and P. J. Wang. 2012. MNS1 is essential for spermiogenesis and motile ciliary functions in mice. PLoS Genet. 8(3): e1002516.

Proteomic analysis of different bovine epithelial cells phenotypes after *Escherichia coli* LPS challenge

Cristian Piras[1*], Yongzhi Guo[2], Alessio Soggiu[1], Viviana Greco[3], Andrea Urbani[3,4], Luigi Bonizzi[1], Patrice Humblot[5] and Paola Roncada[1,6]

[1]Dipartimento di Scienze Veterinarie e Sanità Pubblica, Università degli studi di Milano, Milano, Italy; cristian.piras@unimi.it

[2]Division of Reproduction, Departement of Clinical Sciences,Faculty of Veterinary Medicine and Animal Science, Sweden

[3]Fondazione Santa Lucia – IRCCS, Rome, Italy

[4]Dipartimento di Medicina Sperimentale e Chirurgia, Università degli Studi di Roma 'Tor Vergata', Italy

[5]University of Agricultural Sciences, SLU, Sweden

[6]Istituto Sperimentale Italiano L. Spallanzani, Milano and Spallanzani technologies srl, Italy

Objectives

Infertility in cows represents one of the main problems for milk production resulting in a negative economic impact for dairy industry. It has been demonstrated how fertility rate is decreasing with the increment of milk production suggesting a putative link (Rodriguez-Martinez *et al.*, 2008). One of the main causes seems to be related to post-partum bacterial infections.

Escherichia coli is the most common bacteria associated with cow post partum metritis and endometritis and is related to the disruption of uterine and ovarian function. Its pathogenic effects are induced by the outer-membrane component Lipopolysaccharide (LPS) that is recognized by Toll-like receptors (TLRs) on endometrial cells and leads to secretion of cytokines, chemokines and anti-microbial peptides (Sheldon *et al.*, 2009). Moreover it has been demonstrated that LPS stimulation can modulate the synthesis of PGF2α and PGE2 (Herath *et al.*, 2006).

Strong proliferative properties of LPS have been recently demonstrated in different cell types including bovine endometrial cells. The mechanisms involved in epithelial cells proliferation and the putative link to cattle fertility still has to be elucidated. In this work it has been investigated the differential protein expression of bovine epithelial cells challenged with different LPS concentrations in order to evaluate the mechanisms involved in host-pathogen interaction and in the stimulation of cell proliferation putatively linked to cattle infertility.

Materials and methods

Bovine endometrial (EEC) epithelial cell were challenged with 0, 8 and 16 µg/ml *E. coli* LPS. The variation in cell number among groups (control, epithelial cells challenged with 8 µg/ml LPS, epithelial cells challenged with 16 μg/ ml LPS) was calculated 72 hours after challenge. Cells were collected and pellets deep frozen until use.

The proteomic profile was determined by 2D electrophoresis followed by MALDI TOF-MS and SHOTGUN-MS in controls and LPS-treated cells. About 2×10^6 cells were lysed in 2DE buffer containing 7 M urea, 2 M thiourea, 4% CHAPS, 1% DTT and protease inhibitors (GE Healthcare). 2D electrophoresis was performed using 3-10 IPG strips for isoelectric focusing and SDS PAGE was performed in 12% acrylamide gel. After runs, gels were stained with colloidal Coomassie and digitalized with PharosFX Plus Laser Imaging System (BioRad). Image analysis was performed using Progenesis SameSpots (v4.5; Nonlinear Dynamics, Newcastle, UK). Proteins differentially expressed that showed a p-value lower than 0.05 were taken in consideration for mass spectrometry analysis. Protein spots were manually excised. After reduction and alkylation proteins were digested with a solution of 0.01 μg/μl of porcine trypsin (Promega, Madison, WI) at 37 °C for 16 h. Mass spectra were acquired with an Ultraflex III MALDI-TOF/TOF spectrometer (Bruker-Daltonics) in positive reflectron mode (Turk *et al.*, 2012).

For shotgun proteomics analysis, after protein digestion with trypsin (Promega, Madison, WI, USA), 0.75 μg of digested protein was loaded on a nanoACQUITY UPLC System (Waters Corp., Milford, MA, USA) coupled to a Q-Tof Premier mass spectrometer (Waters Corp., Manchester, UK).

Continuum LC-MS data were processed using ProteinLynx GlobalServer v2.4 (PLGS, Waters Corp. Manchester, UK).

Protein identification was performed using the UniProtKB/Swiss-Prot bovine database (Soggiu *et al.*, 2013).

Results and discussion

As found previously, cell growth rate was steadily increased for cells exposed to 8 μg/ml when compared to controls (range +20% to +40%; $P<0.001$).

Stimulation of cell growth following exposure to 8 μg/ml LPS was associated to significant differential protein expression. Collected data about different growth profile after LPS stimulation are resumed in Figure 1.

From preliminary results it was highlighted an increased expression of Interferon-induced dynamin-like GTPase, Protein disulfide-isomerase A3, and of transketolase. Among other proteins up-regulated after LPS challenge there were found many proteins involved in stress response such as chaperones and proteins active against oxidative stress. Figure 2 shows the differential expression profile of proteins obtained respectively through 2D electrophoresis and mass spectrometry.

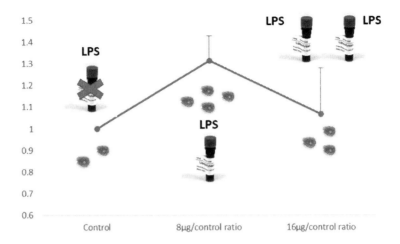

Figure 1. Growth profile of epithelial cells after E. coli *LPS challenge*

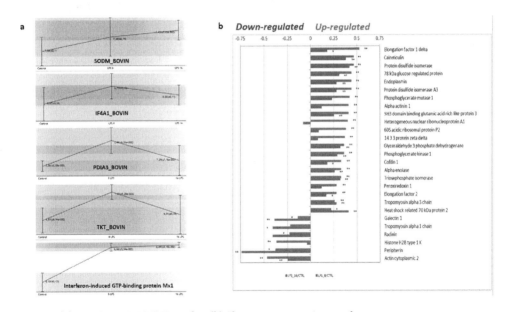

Figure 2. (a) 2DE/MALDI-TOF results; (b) Shotgun proteomics results.

Among obtained results it has to be highlighted the over-expression of interferon-induced dynamin-like GTPase. This protein was found to be directly linked to host defence and over-expressed after LPS stimulation.

These preliminary results indicate that LPS-induced proliferation is associated with changes in protein profiles associated to resistance to pathogens, general metabolism, membrane activity and regulation of DNA methylation. These results in protein expression and metabolisms could be related to cattle infertility and next step will be to validate these results on a larger number of animals.

Acknowledgements

Work supported by EU7FP prolific project (PR_ST).

References

Herath, S., Dobson, H., Bryant, C. and Sheldon, I., 2006. Use of the cow as a large animal model of uterine infection and immunity. Journal of reproductive immunology 69: 13-22.

Rodriguez-Martinez, H., Hultgren, J., Båge, R., Bergqvist, A.-S., Svensson, C., Bergsten, C., Lidfors, L., Gunnarsson, S., Algers, B. and Emanuelson, U., 2008. Reproductive performance in high-producing dairy cows.

Sheldon, I.M., Cronin, J., Goetze, L., Donofrio, G. and Schuberth, H.-J., 2009. Defining postpartum uterine disease and the mechanisms of infection and immunity in the female reproductive tract in cattle. Biology of reproduction 81: 1025-1032.

Soggiu, A., Piras, C., Hussein, H.A., De Canio, M., Gaviraghi, A., Galli, A., Urbani, A., Bonizzi, L. and Roncada, P., 2013. Unravelling the bull fertility proteome. Molecular BioSystems 9: 1188-1195.

Turk, R., Piras, C., Kovačić, M., Samardžija, M., Ahmed, H., De Canio, M., Urbani, A., Meštrić, Z.F., Soggiu, A. and Bonizzi, L., 2012. Proteomics of inflammatory and oxidative stress response in cows with subclinical and clinical mastitis. Journal of proteomics 75: 4412-4428.

Proteome adaptation of boar semen during storage

Viviana Greco[1,2]*, Blanka Premrov Bajuk[3], AlessioSoggiu[4], Cristian Piras[4], Petra Zrimšek[5], Maja Zakošek Pipan[5], Luigi Bonizzi[4], Andrea Urbani[1,2] and Paola Roncada[4,6]

[1]*Department of Experimental Medicine and Surgery, University of Rome 'Tor Vergata', Rome, Italy; vivianagreco82@yahoo.it*

[2]*Proteomics and Metabonomics Laboratory, Santa Lucia Foundation-IRCCS, Rome, Italy,*

[3]*Institute of Physiology, Pharmacology and Toxicology, Veterinary Faculty, University of Ljubljana, Slovenia*

[4]*Dipartimento di Scienze Veterinarie e Sanità Pubblica (DIVET), Università di Milano, Milano, Italy*

[5]*Clinic for Reproduction and Horses, Veterinary Faculty, University of Ljubljana, Slovenia*

[6]*Istituto Sperimentale L. Spallanzani, Milano, Italy*

Introduction

Artificial insemination (AI) with extended semen offers many benefits to the swine industry through improving biosecurity and access to high-quality genetic material (Estrada *et al.*, 2014). Modern boar industry worldwide is based on the use of AI with extended semen cooled at 15-20 °C for 1-5 days (Johnson *et al.*, 2000). The semen storage represents a key point in this process and for this reason it is necessary evaluate all the steps for proper sample collection. Conventional evaluation of semen quality is generally based on measure of sperm concentration, viability, motility and morphology (Bonet *et al.*, 2012; Dyck *et al.*, 2011). Therefore, these parameters do not consider the functionality of spermatozoa, an important factor related to reproductive performance. Semen preservation could generate several negative effects on spermatozoa due to dilution, change of microenvironment, chilling and ageing (Waberski *et al.*, 2011). In particular, boar semen, probably because of the low cholesterol/phospholipid ratio of cell membrane (Dubé *et al.*, 2004), is especially sensitive to cold shock compared to other domestic animals (Benson *et al.*, 2012). Even if the use of pooled semen samples and the relatively high number of spermatozoa for the commercial AI compensate for a reduced fertility, the study of the alteration of sperm cells during room temperature storage could be of high scientific impact. Moreover the comprehensive understanding of the proteome of a sensitive model such as boar semen could be useful to elucidate storage associated changes of protein expression. The aim of this study was to investigate whether and how liquid storage at room temperature for three days causes an alteration in the proteomic profile of stored spermatozoa. In fact proteomics analysis of sperm represents a valuable tool to identify protein expression changes related to fertility (De Canio *et al.*, 2014; Soggiu *et al.*, 2013). The identified differently expressed proteins could be used as biomarkers of semen quality and survival after short-termed storage and this study could contribute to evaluate the putative additeves useful to improve sperm ferility after storage.

Materials and methods

Samples of fresh and stored (15-17 °C for 3 days) semen were analysed.

Ten semen samples from eight 12- to 24-month-old boars of various breeds that are routinely used at the local AI centre were collected using glove-hand technique. Gel, dust and bristles were filtered out by a semen collecting flask. After storage, characteristics of liquid stored semen samples were evaluated in the same way as for fresh samples. The difference between fresh and stored semen samples was tested using t-test. $P<0.05$ was considered as significant. SigmaStat 3.5 (SYSTAT Software Inc.) software was used for all tests.

For proteomic analysis, eight pools of protein extracts (4 from fresh and 4 from stored semen) were prepared and each pool included 5 protein extracts (a total of approximately 95 µg of protein), randomly prepared from 10 samples. The protein profile of samples was determined using two-dimensional electrophoresis. After run, gels were stained with colloidal Coomassie Blue G250 solution and de-stained in MilliQ water.

Images were acquired using ImageScanner III (GE Healthcare) and analysed by ProgenesisSameSpotv. 4.5 software (Nonlinear Dynamics). Spot detection and normalisation were performed using the automated tools of the software. Principal component analysis (PCA) was elaborated using the Stat module of this software. The One-way Anova test was used to compare the protein expression profiles of fresh semen to semen after 3 days of storage. P values less than 0.05 were reported as statistically significant. The differencially expressed spots were excised form gels and analysed with an Ultraflex III MALDI-TOF/TOF spectrometer (Bruker-Daltonics) for protein identification.

Results and discussions

Comparative quantitative image analysis of 2D gels by Progenesis Same Spot revealed three spots that significantly differ in liquid stored samples relative to fresh sperm.

Figure 1 shows differentially expressed spots that have been cut and analysed through mass spectrometry.

The spot 241 has been identified as ATP citrate lyase that is involved in replacement of oxidative-stress damaged fatty acid biosynthesis. The spot 568 has been identified as Chaperonin containing TCP1 plays a key role in sperm-oocyte interaction. The spot 629 correspond to Cytosolic non specific dipeptidase that has been documented to be involved in human male fertility.

Interestingly, this experiment shows that ATP citrate lyase was found to be upregulated while Chaperonin containing TCP1 and Cytosolic non specific dipeptidase were found to be upregulated in liquid stored samples compared to fresh sperm. As documented from our results, oxidative stress during long term sperm cells storage is one of major reason of cells death because of the progressive depletion of fatty acids for cellular structures. Cellular response to counteract this lack in the increased synthesis of fatty acids by enzymes such as ATP citrate lyase.

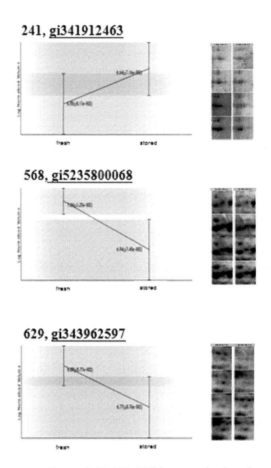

Figure 1. Differentially expressed spots (241, 568, 629) between fresh and stored semen samples. All the represented spots had P<0.05.

Acknowledgements

Author are grateful to the COST ACTION FA1002 Farm Animal Proteomics for the support of the Short Term Mission.

References

Benson, J.D., Woods, E.J., Walters, E.M. and Critser, J.K., 2012. The cryobiology of spermatozoa. Theriogenology 78: 1682-1699.

Bonet, S., Briz, M. and Yeste, M., 2012. A proper assessment of boar sperm function may not only require conventional analyses but also others focused on molecular markers of epididymal maturation. Reproduction in Domestic Animals 47: 52-64.

De Canio, M., Soggiu, A., Piras, C., Bonizzi, L., Galli, A., Urbani, A. and Roncada, P., 2014. Differential protein profile in sexed bovine semen: shotgun proteomics investigation. Molecular BioSystems 10: 1264-1271.

Dubé, C., Beaulieu, M., Reyes-Moreno, C., Guillemette, C. and Bailey, J.L., 2004. Boar sperm storage capacity of BTS and Androhep Plus: viability, motility, capacitation, and tyrosine phosphorylation. Theriogenology 62: 874-886.

Dyck, M., Foxcroft, G., Novak, S., Ruiz-Sanchez, A., Patterson, J. and Dixon, W., 2011. Biological markers of boar fertility. Reproduction in Domestic Animals 46: 55-58.

Estrada, E., Rodríguez-Gil, J., Rocha, L., Balasch, S., Bonet, S. and Yeste, M., 2014. Supplementing cryopreservation media with reduced glutathione increases fertility and prolificacy of sows inseminated with frozen-thawed boar semen. Andrology 2: 88-99.

Johnson, L., Weitze, K., Fiser, P. and Maxwell, W., 2000. Storage of boar semen. Animal reproduction science 62: 143-172.

Soggiu, A., Piras, C., Hussein, H.A., De Canio, M., Gaviraghi, A., Galli, A., Urbani, A., Bonizzi, L. and Roncada, P., 2013. Unravelling the bull fertility proteome. Molecular BioSystems 9: 1188-1195.

Waberski, D., Henning, H. and Petrunkina, A., 2011. Assessment of storage effects in liquid preserved boar semen. Reproduction in Domestic Animals 46: 45-48.

Metaproteomic analysis, application to water quality assessment in aquaculture

Jacob Kuruvilla[1], Binu Mathew[1], Gabriela Danielsson[2], Alessio Soggiu[3], Paola Roncada[3] and Susana Cristobal[1,4*]

[1]Department of Clinical and Experimental Medicine, Cell biology, Faculty of Health Science, Linköping University, Linköping, Sweden; susana.cristobal@liu.se

[2]Department of Biochemistry and Biophysics, Arrhenius laboratories, Stockholm University, Stockholm, Sweden

[3]Dipartimento di Scienze Veterinarie e Sanità Pubblica (DIVET), Università degli Studi di Milano, Milano, Italy

[4]IKERBASQUE and Department of Physiology, Faculty of Medicine and Dentistry of University of Basque Country, Leioa, Spain

Current metaproteomics techniques could provide a wide resolution of proteins from a complex ecosystem biological or environmental. The qualitative and quantitative metaproteomic analysis provides a functional overview over several level of complexity: at the community level, how the communities respond to the exposure, stress or disease; at the molecular level, which protein families have been expressed; and at the biochemical level, which enzymatic reaction or macromolecular machines have acquired a relevant role to maintain the community in equilibrium. Thus, metaproteomics analysis could offer a molecular snap shot of the eukaryote or prokaryote communities at a certain time and place.

This alternative methodology has been applied to analyze the functional role of the microbial community in cheese (Soggiu et al., 2014), in bioremediation, characterizing how microbial species could remediate toxic metal contamination in soils and ground waters (Chourey et al., 2013), in carbon cycling, defining the role that microbial species have in the flow of carbon in a given ecosystem (Bastida et al., 2014), in bioenergy, by understanding how microbial species might help convert cellulosic material to biofuels (Lu et al., 2014) or most extensively in human health and personal medicine, identifying how microbial species impact/control disease vs health in organs (Jones et al., 2014; Verberkmoes et al., 2009).

However, this alternative method has not yet explored to implement traditional environmental assessment. Environmental assessment has traditionally focussed on evaluating the environmental quality by measuring abiotic components. The fundamental goal of a metaproteomic environmental assessment is to deliver a robust evaluation the environmental impact at population, individual, cellular and molecular level from one single, comprehensive and cost-efficient methodology. Considering the importance of water quality control in aquaculture, we tested if a metaproteomic-based assessment can estimate: (1) variation in the abundance of protein families; (2) changes in biodiversity; and (3) changes in the equilibrium of a marine soil ecosystem after exposure to a stressor.

We have performed a laboratory experiment building a microcosm based on the minimal number of species that are required to mimic the structure and function of the coastal Baltic Sea community. This microcosm consisted of macroalga (*Ceramium tenuicorne*), mussels (*Mytilus edulis trossulus*), amphipods (*Gammarus spp.*), water and sediment. The microcosms have been exposed to the human pharmaceutical (propranolol) and in combination with changes in salinity. The sampled organisms and sediment were distributed among 15 glass aquaria and acclimated for seven days in a climate chamber before exposure. The same conditions persisted throughout the experimental period. Fresh seawater (salinity 6.52±se 0.0075 psu) was continuously added into each separate aquaria via a flow through system (Oskarsson *et al.*, 2014). The sediment were selected for <3 mm and exposed for a 6 week mesocosm model with (1) the environmental relevant exposure of 100 µg/l of pharmaceutical propranolol and a high level exposure of 1000 µg/l; and (2) the additional stressor of lowered salinity mimicking the consequence of global warming with +2 °C combined with the pharmaceutical exposure in (1). The three species of algae, crustacean and mollusc in the microcosm represents a small scale ecosystem with an embedded food chain and sediment selected for <3 mm biota. The experiment concentrations of the exposure levels has been quantified to average 109±12 respectively 1,092±111 µg/l (Oskarsson *et al.*, 2014).

The proteins from the sediment were extracted with SDS, separated by SDS-PAGE. The gel was divided in 15 slides and the protein mixtures were in-gel trypsin digested. The peptides were analysed on a linear trapping quadruple-Orbitrap-MS. The LC-MS (2)-spectra were then searched using a combination of search engines against NCBI non-redundant repository database and computational analysis. Metaproteomics data would provide a representation of the populations at different taxonomic levels on the bases of the taxon-specificity of a tryptic peptide list.

Our results indicate that applying a metaproteomic analysis based in the estimation of the relative abundance of protein families from a complex community could be correlated to both changes in the biodiversity and the molecular responses of the communities. Therefore it could be apply to water quality control in aquaculture, in particular. In general, the environmental problems from the past, the environmental risks from today are urgent to identify, quantify and predict. The risk assessment methodologies and field biomonitoring cannot apply to the complexity of our emergent environmental risks and pollutants. Therefore, solving the environmental assessment challenge with one single methodology will contribute our common search for sustainable development.

References

Bastida, F., Hernandez, T., Garcia, C., Metaproteomics of soils from semiarid environment: Functional and phylogenetic information obtained with different protein extraction methods. Journal of proteomics 2014, 101, 31-42.

Chourey, K., Nissen, S., Vishnivetskaya, T., Shah, M., *et al.*, Environmental proteomics reveals early microbial community responses to biostimulation at a uranium- and nitrate-contaminated site. Proteomics 2013, 13, 2921-2930.

Jones, M. L., Martoni, C. J., Ganopolsky, J. G., Labbe, A., Prakash, S., The human microbiome and bile acid metabolism: dysbiosis, dysmetabolism, disease and intervention. Expert opinion on biological therapy 2014, 14, 467-482.

Lu, F., Bize, A., Guillot, A., Monnet, V., *et al.*, Metaproteomics of cellulose methanisation under thermophilic conditions reveals a surprisingly high proteolytic activity. The ISME journal 2014, 8, 88-102.

Oskarsson, H., Eriksson Wiklund, A. K., Thorsen, G., Danielsson, G., Kumblad, L., Community interactions modify the effects of pharmaceutical exposure: a microcosm study on responses to propranolol in Baltic Sea coastal organisms. PloS one 2014, 9, e93774.

Soggiu, A., Piras, C., Levi Mortera, S., Brasca, M., Urbani, A., Bonizzi, L. and Roncada, P., 2014. The Grana Padano microbiota: insights from experimental caseification. In: De Almeida, A., *et al.* (ed.), 2014. Farm animal proteomics 2014 – Proceedings of the 5th Management Committee Meeting and 4th Meeting of Working Groups 1,2 & 3 of COST Action FA 1002, Wageningen Academic Publishers, Wageningen, the Netherlands, pp. 59-61.

Verberkmoes, N. C., Russell, A. L., Shah, M., Godzik, A., *et al.*, Shotgun metaproteomics of the human distal gut microbiota. The ISME journal 2009, 3, 179-189.

Fat&MuscleDB: integrating 'omics' data from adipose tissue and muscle

Jérémy Tournayre[1,2]*, Isabelle Cassar-Malek[1,2], Matthieu Reichstadt[1,2], Brigitte Picard[1,2], Nicolas Kaspric[1,2] and Muriel Bonnet[1,2]*

[1]*INRA, UMR1213 Herbivores, 63122 Saint-Genès-Champanelle, France*
[2]*Clermont Université, VetAgro Sup, UMR1213 Herbivores, BP 10448, 63000, Clermont-Ferrand, France; jeremy.tournayre@clermont.inra.fr; muriel.bonnet@clermont.inra.fr*

Introduction

Increasing the amount of meat while preserving a high quality is an economic challenge for the beef industry. From a biological point of view, this aim depends on the relative amounts of muscle and adipose tissues that determines the lean-to-fat ratio. Understanding how rearing practices affect the lean-to-fat ratio has implications for the quality and value of the carcass and the meat from cattle. It had thus motivated vast amounts of transcriptomic and proteomic studies to identify which genes and proteins control tissue growth and physiology. We hypothesize that integration of available published data through a computational strategy will foster discoveries of the genes or proteins that are systematically related to stages of the development of adipose tissue and muscle, as candidate markers of potential tissue growth. Moreover, the assembly of contextualized knowledge will improve our understanding of the interactions between adipose tissue and muscle. Indeed, most of the 'omics' studies questioned the growth of either muscles or adipose tissues in cattle depending on intrinsic (genotype, sex) or environmental (nutrition) factors (Picard *et al.* 2011). They were never merged to question the developmental and functional links between muscle and adipose tissue. These links are suggested by the successive waves of growth of muscle and adipose tissue as well as the wide plasticity of body composition depending on age or genotype in cattle (Bonnet *et al.*, 2010). Currently, some functional links are rarely reported in monogastrics thanks to *in vitro* studies (Romacho *et al.*, 2014).

To gather available data, we designed the Fat&MuscleDB, an online database of transcriptomic and proteomic data from skeletal muscle and adipose tissue. Because muscular and adipose masses grow by hyperplasia during foetal or early post-natal life in cattle as well as by hypertrophy, we integrated data related to these mechanisms both *in vivo* and *in vitro*, from ruminants but also from models species or cell lines experiments. We defined criteria relative to the experiments and the stages of adipose tissue and muscle growth in order to record and classify data. Our final objective is to aggregate all references that share the same criterion related to a stage of adipose tissue and muscle growth to help biologists to find candidate proteins to be biomarkers of the lean-to-fat ratio.

Methods

A database of transcripts and proteins (with their abundance values) dealing with the growth of adipose and muscle tissue has been built from Pubmed and Gene Expression Omnibus (GEO) on the National Center for Biotechnology Information (NCBI) and the Web of Knowledge. To select contextualized articles or dataset, keywords were defined by experts in fat and muscle growth and submitted via an automatic procedure. Each retrieved publication (with its supplementary data) was analysed manually to check for results and table(s) reporting transcripts or proteins differentially abundant during muscle and/or fat growth. The extraction of tables from a PDF was simplified by the use of a tool that has been implemented. We used 'pdftotext' command to convert PDF to text and then to a tabbed file. Transcriptomics results of GEO were analysed to determine transcripts differentially expressed or not during muscle and/or fat growth. Curators had chosen, among a GEO Series (GSE), GEO Samples (GSM) to submit to analyse in the context of adipose tissue and muscle growths. Transcriptomics results were then recovered and processed using a tool developed in R software with libraries GEOquery (Davis and Meltzer, 2007) and Anapuce (http://cran.rproject. org/web/packages/anapuce). Anapuce was chosen for its specific functions ('DiffAnalysis' and 'DiffAnalysis.unpaired') for the differential analysis between two groups of microarray data. Data were normalized when necessary through 'normalisation' function, spots for a same transcript and technical replicates were averaged and the variance mixture was calculated with 'DiffAnalysis' or 'DiffAnalysis.unpaired' functions for paired data or unpaired data, respectively. We chose a p-value of 0.15 for the variance because of the stringency of this test that strongly reduced the number of regulated transcripts. Each of the data analysed by these methods was verified by curators and annotated as 'Analysed by Fat&MuscleDB'. In contrast, data recorded as they were published were annotated as 'Not analysed by Fat&MuscleDB'. Data were recorded with information given by the authors regarding the experimental protocols in order to allow investigators to conduct researches and aggregations in Fat&MuscleDB.

In order to record, classify and aggregate data, we defined 73 criteria on experiments and stages of adipose tissue and muscle growth. The aggregation of data implies to have a unique identifier for a gene or a protein. We have chosen to convert all names or identifiers in UniProt accession from UniProt which is a protein database regularly updated and curated by experts. To do this, each name or identifiers was sent on the search engine of UniProt or NCBI in order to get the UniProt accession for the species investigated, otherwise in any species for not losing information. By this way, the aggregation process produced a table that merged results from all experiments sharing a same criterion, and all transcripts or proteins were referred by a UniProt accession. Each accession was counted once by reference in order to identify the redundancy of identification of a gene or a protein.

Results and discussion

Three lists of keywords have been created to seek references from PubMed, GEO and Web of Knowledge in order to establish a reference database. One list described cellular features of adipose and muscle tissues (hypermuscularity, high adiposity, myoblast, adipocyte...), another characterizes species and cell lines (bovine, human, 3T3-L1, C2C12) and the latest relates methods (transcriptomics, proteomics). Association of keywords between these lists has generated approximately 27,000 combinations that were automatically launched on public databases. Moreover, in order to refine the search, negative keywords have been added to each association. Following this search, we found 15,552 publications and 2,312 GEO datasets. Right now, 2,366 publications and 928 GEO datasets does not contain information that interests us contrary to 185 publications and 139 GEO datasets must be processed to be inserted into the database. Among those to be treated, we recovered 164 side-by-side comparisons from 26 publications and 18 GEO datasets in order to identify genes or proteins with an stable, increased or decreased abundance. We continue to feed Fat&MuscleDB, there are still over 250 references on all species to treat and over 12,000 references to analyse.

We report the example of the integration of data related to the criterion 'Muscular hypertrophy from genetic origin' (Figure 1). By clicking on the 'Aggregation' button, investigators have to select this criterion among others and to select other settings to submit the integration. Currently this integration uses data from six references analysed by twelve comparisons. Thus, UniProt accession lists are built up of transcripts and proteins according to their abundance. After the integration of the data by the aggregation process we identified 13 proteins and 357 transcripts with a decreased abundance, 3 proteins and 286 transcripts with an increased abundance and 17,088 transcripts with a stable abundance that are related to 'Muscular hypertrophy from genetic origin'. This list will be submitted for data mining using informatics resources such as ProteINSIDE that we have developed (Kaspric *et al.*, 2014), to identify central or relevant genes or proteins related to muscular hypertrophy. Aggregation can be achieved for the 72 other criteria that we have defined to depict all the stages of adipose tissue and muscle growth.

The time devoted to analysing references is substantial, so references dealing with ruminants were analysed in priority. However, because proteins and transcripts underlying fat and muscle growth can be the same in several species, we will complete our database with data from human and cells lines. Fat&MuscleDB provides an innovative and interactive resource that connects beef industry interests with the new biological discoveries in 'omics' research. This computational strategy will catalyze the assembly of knowledge about muscle and adipose tissue growth, and foster the discovery of candidate biomarkers of the lean-to-fat-ratio. This would also minimize unnecessary redundancy in research efforts.

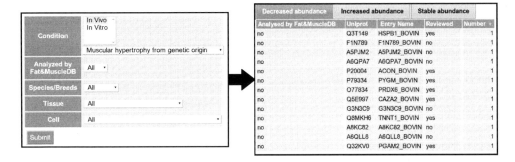

Figure 1. Integration of 'omics' data relative to 'Muscular hypertrophy from genetic origin' on Fat&MuscleDB. (Left) The form that allows selecting references to aggregate according to different settings: in vitro or in vivo experiment, if the data were analysed by us or recovered according to author interpretation, breeds, species, tissues and lines cell. (Right) Extract of the list of proteins whose abundance decreases obtained by aggregating data for 'Muscular hypertrophy from genetic origin'. Each line corresponds to a protein with multiple indications: if it was recovered by us ('Yes') or recovered according to author interpretation ('No'), its UniProt accession, its Entry name, if it has been reviewed by UniProt consortium or not, and the number of times it is aggregated (a maximum of one by reference). Here, all proteins were recorded by only one reference.

Acknowledgements

This study was supported by the regional council of Auvergne in France through the regional information system Lifegrid.

References

Bonnet, M., Cassar-Malek, I., Chilliard, Y. and Picard, B., 2010. Ontogenesis of muscle and adipose tissues and their interactions in ruminants and other species. Animal 4: 1093-1109.

Davis, S. and Meltzer, P.S., 2007. GEOquery: a bridge between the Gene Expression Omnibus (GEO) and BioConductor. Bioinformatics 23: 1846-1847.

Kaspric, 2014. Proteomics data from ruminants easily investigated using ProteINSIDE. In: De Almeida, A., *et al.* (ed.), 2014. Farm animal proteomics 2014 – Proceedings of the 5th Management Committee Meeting and 4th Meeting of Working Groups 1,2 & 3 of COST Action FA 1002, Wageningen Academic Publishers, Wageningen, the Netherlands, pp. 283-287.

Picard, B., Cassar-Malek, I., Guillemin, N. and Bonnet, M., 2011. Quest for Novel Muscle Pathway Biomarkers by Proteomics in Beef Production. In: M. Moo-Young (ed.), Comprehensive Biotechnology (Second Edition). Academic Press, Burlington, pp. 395-405.

Romacho, T., Elsen, M., Rohrborn, D. and Eckel, J., 2014. Adipose tissue and its role in organ crosstalk. Acta Physiol (Oxf) 210: 733-753.

Developing new methods for PTM mapping of *Ehrlichia ruminantium* proteins

Isabel Marcelino[1,2,3*], Núria Colomé-Calls[4], Rita Laires[2], Thierry Lefrançois[5], Anna Bassols[6], Nathalie Vachiéry[3], Ana V. Coelho[2] and Francesc Canals[4]

[1]Instituto de Biotecnologia Experimental e Tecnológica (IBET), Apartado 12, 2780-901 Oeiras, Portugal; isabel_m31@hotmail.com

[2]Instituto de Tecnologia Química e Biológica, Universidade Nova de Lisboa (ITQB), Av. da República, 2780-157 Oeiras, Portugal

[3]Centre de coopération Internationale en Recherche Agronomique pour le Développement (CIRAD), UMR CMAEE, 97170 Petit-Bourg, Guadeloupe, FWI

[4]Proteomics Laboratory, Vall Hebron Institute of Oncology (VHIO), Institut de Recerca Hospital Univ. Vall Hebron Collserola Building, Pg. Vall d'Hebron, 119-129, 08035 Barcelona, Spain

[5]Centre de coopération Internationale en Recherche Agronomique pour le Développement (CIRAD), UMR CMAEE, 34398 Montpellier, France

[6]Departament de Bioquímica i Biologia Molecular, Servei de Bioquímica Clínica Veterinària, Facultat de veterinària, Universitat Autònoma de Barcelona (UAB), 08193 Cerdanyola del Vallès, Spain

Objectives

Ehrlichia ruminantium (ER) is an obligate intracellular bacterium of the *Rickettsiales* order that causes Heartwater, a fatal tick-borne disease in ruminants, posing important economic constraints to livestock production in sub-Saharan Africa and in two Caribbean islands, where it threatens the American mainland. Several vaccine candidates (inactivated, attenuated and recombinant) are under evaluation, but the development of a fully effective vaccine has been hindered by the high antigenic diversity of strains and, specially, to the lack of knowledge on ER biology and pathogenesis (Vachiery *et al.*, 2013). The mechanisms of invasion and development of ER in host cell are also unknown.

Recently, our group showed that, among the most abundant proteins expressed in the ERGardel strain, 25% had isoforms (Marcelino *et al.*, 2012). New data on whole 2-DE proteomic profiling of ER elementary bodies from virulent and attenuated strains revealed that isoforms could account up to 85% of the identified proteins (unpublished data).

In pathogenic bacteria, post-translationally modified proteins can promote bacterial survival, replication, and evasion from the host immune system. PTMs are also known to have a critical impact in vaccine efficacy. Therefore, complete PTM mapping of ER proteins is highly desirable. For this, we developed new enrichment protocols for ER glycoproteins and phosphopeptides followed by mass spectrometry analysis for protein identification.

Materials and methods

ER biological triplicates (Gardel strain, from Guadeloupe) were produced in bovine aortic endothelial cells (BAE) as described elsewhere (Marcelino *et al.* 2005). At cell lysis, infectious elementary bodies (EBs) were harvested, purified with a multistep centrifugation process (Marcelino *et al.* 2007) and stored in SPG buffer at -80 °C until use. ER protein extracts were prepared using a NP40-SDS containing buffer and several cycles of sonication. The total amount of protein per ER sample was quantified using the RC-DC protein assay kit (Biorad) according to the manufacturer's instruction. To identify ER phosphoproteins, a protocol developed by Thingholm and Larsen was used (Thingholm *et al.* 2006). Briefly, tryptic peptides prepared using a filter-aided sample preparation (Wisniewski *et al.* 2009) were diluted in glycolic acid and passed several times through the TiO2 columns. Phosphopeptides were then eluted with NH_4OH and sample pH was adjusted with formic acid and analyzed by mass spectrometry. To identify ER glycoproteins, ER samples and ULH resin were prepared according the manufacturer's instructions. Briefly, proteins were oxidized and incubated overnight with the ULH resin at RT. Proteins bound to the resin were then digested and the tryptic peptides were used for to MS analysis. All LC-MS/MS analyses were performed on a Maxis Impact Q-TOF spectrometer (Bruker, Bremen) coupled to a nano-HPLC system (Proxeon, Denmark). Proteins were identified using Mascot (Matrix Science, London UK).

Results and discussion

To evaluate the efficacy of TiO2 columns for phosphopeptide enrichment, we first performed a search on ProteinScape using a *Bos taurus* database, since bovine endothelial cells are known to have phosphoproteins and are unavoidable contaminants in ER samples. The results showed that despite the detection of some unspecific proteins, it was possible to detect bovine phosphoproteins (such as HSP 27, Alpha-2-HS-glycoprotein, Cysteine string protein beta) with reasonable MASCOT scores (between 30-40) and good MS/MS fragmentation spectra. These results validated the use of TiO2 columns for the enrichment of phosphopeptide in these samples. Still, when MS/MS data were processed using an ER database, only one protein (Pyruvate phosphate dikinase) was detected. Although the reasons for this are not yet clear, this could be due to (1) a low amount of ER proteins initially used for phosphopeptide enrichment; (2) to a real low amount of phosphoproteins in ER; or (3) that the ER phosphopeptides were hidden by the bovine phosphopeptides that are more abundant and have a better ionization.

Selective enrichment of glycoproteins was performed using Ultralink hydrazide (ULH) resin. Since this resin was used for the first time to detect ER glycoproteins, we first tested the specificity and efficacy of the resin. For this, oxidized and non-oxidized ER protein extracts were used. The preliminary results clearly indicated that the protocol was highly specific for oxidized proteins (and therefore glycosylated proteins), with less than 1% false positive for bovine contaminants and 5% for ER proteins. Preliminary assays revealed that more than 50% of ER proteome would be composed of glycoproteins, which could then explain the high number of isoforms detected in previous 2DE experiments. This corroborates with the preliminary results obtained by Postigo and co-workers regarding the glycosylation of MAP1 protein isoforms.

Conclusions and future work

This work shows that the methods developed herein allow the enrichment of ER glycoproteins and phosphopeptides, for further identification of post-translationally modified proteins using mass spectrometry. We are currently applying these protocols to other ER strains that present different levels of virulence. We then expect to (1) provide a global overview of the post-translationally modified proteins in ER and their respective PTMs; and (2) identify ER proteins possibly involved in bacteria antigenic diversity and ER pathogenesis, with potential interest for vaccine development. As these protocols were also validated for host endothelial cells protein extract, they will be useful to detect post-translationally modified proteins in the host cell during the infection process.

Acknowledgements

Authors acknowledge funding from the EU projects COST action FA-1002 (for networking opportunities) and FEDER 2007-2013 (FED 1/1.4-30305) and also the project ER-TRANSPROT (PTDC/CVT/114118/2009), post-doc grant SFRH/ BPD/ 45978/ 2008 (I. Marcelino) all financed by Fundação para a Ciência e a Tecnologia (Lisboa, Portugal).

References

Marcelino, I., A. M. de Almeida, C. Brito, D. F. Meyer, M. Barreto, C. Sheikboudou, C. F. Franco, *et al.* 2012. Proteomic analyses of Ehrlichia ruminantium highlight differential expression of MAP1-family proteins. *Vet Microbiol* 156 (3-4): 305-14.

Marcelino, I., N. Vachiery, A. I. Amaral, A. Roldao, T. Lefrancois, M. J. Carrondo, P. M. Alves, and D. Martinez. 2007. Effect of the purification process and the storage conditions on the efficacy of an inactivated vaccine against heartwater. *Vaccine* 25 (26): 4903-13.

Marcelino, I., C. Verissimo, M. F. Sousa, M. J. Carrondo, and P. M. Alves. 2005. Characterization of Ehrlichia ruminantium replication and release kinetics in endothelial cell cultures. *Vet Microbiol* 110 (1-2): 87-96.

Thingholm, T. E., T. J. Jorgensen, O. N. Jensen, and M. R. Larsen. 2006. 'Highly selective enrichment of phosphorylated peptides using titanium dioxide.' In *Nat Protoc*, 1929-35. England.

Vachiery, N., I. Marcelino, D. Martinez, and T. Lefrancois. 2013. 'Opportunities in diagnostic and vaccine approaches to mitigate potential heartwater spreading and impact on the american mainland.' In *Dev Biol (Basel)*, 191-200. Switzerland.

Wisniewski, J. R., A. Zougman, N. Nagaraj, and M. Mann. 2009. 'Universal sample preparation method for proteome analysis.' In *Nat Methods*, 359-62. United States.

Integrative label-free quantitative proteomics study in mastitis

Manikhandan A.V. Mudaliar[1]*, Funmilola C. Thomas[2], Mark McLaughlin[3], Richard Burchmore[1], Pawel Herzyk[1], P. David Eckersall[2] and Ruth Zadoks[2]
[1]Glasgow Polyomics, University of Glasgow, United Kingdom;
manikhandan.mudaliar@glasgow.ac.uk
[2]Institute of Biodiversity, Animal Health and Comparative Medicine, University of Glasgow, United Kingdom
[3]School of Veterinary Medicine, University of Glasgow, United Kingdom

Introduction

Bovine mastitis involves inflammation of the udder, and is an important disease affecting dairy cattle. Mastitis often develops as a sequel to intra-mammary infections, and can be caused by a broad variety of bacterial species. The presence in milk of both immune and pathogen factors can influence dairy production and animal welfare, but make milk a natural substrate to study host-pathogen interactions and in which to identify biomarkers for mastitis. Recent improvements in high-throughput proteomic technologies have enabled characterization of the bovine milk proteome (Boehmer, 2011; Eckersall *et al.*, 2012). We present a novel approach to assess quantitative changes in the proteome and metabolome in bovine milk during the course of a *Streptococcus uberis* infection.

Materials and methods

Milk samples for this study were collected from an intra-mammary challenge experiment with a putatively host-adapted strain of *Streptococcus uberis* (Tassi *et al.*, 2013). Briefly, six clinically healthy Holstein cows with no history of clinical mastitis were intramammarily challenged with *Streptococcus uberis*. Milk samples were collected at different time intervals, and samples from six selected time points (0, 36, 42, 57, 81 and 312 hours post-challenge) were used to generate quantitative label-free proteomics data and quantitative untargeted metabolomics data.

Proteomics data generation and analysis

The milk samples were centrifuged at 150,000×*g* to deplete caseins. Whey proteins were precipitated, re-suspended at a defined concentration and digested with trypsin for proteomic analysis. MS/MS data acquisition was performed using Thermo Scientific Orbitrap Elite mass spectrometer coupled with Dionex UltiMate 3000 RSLCnano in-line HPLC using reversed-phase liquid chromatography on a 120-minute gradient. Peptide identifications, protein assignments and label-free quantifications were performed using MaxQuant software coupled with Andromeda search engine (Cox and Mann, 2008). Differential expression analysis using ANOVA was performed in Partek Genomic

Suite. Perseus software was used for performing data transformation, principal component analysis (PCA) and generating plots. NCBInr *Bos taurus* protein sequences were used in the Andromeda search engine to identify proteins.

Metabolomics data generation and analysis

Using aliquots of the same milk samples assessed by proteomics, metabolites were extracted in 1:3 mixture of chloroform and methanol. LC-MS analysis was performed in both negative and positive ionisation modes using a Thermo Scientific Exactive mass spectrometer coupled with a Dionex UltiMate 3000 RSLCnano in-line HPLC. The separation of metabolites was performed using a ZIC-pHILIC column in a 25-minute gradient. The metabolomics data was analysed using Ideom software (Creek *et al.*, 2012) and R statistical computing software.

Results and discussions

Label-free quantitative proteomics

LC-MS/MS proteomics data was generated from all 36 samples (6 samples per time point) at the six selected time points (0, 36, 42, 57, 81 and 312 hours post-challenge). The quality of the raw data was examined by generating total ion current chromatograms and base peak chromatograms for every sample in the proteomics dataset. The dataset showed overall consistency with very little retention time drift. We identified and quantified 2,399 peptides in the dataset and assigned them to 558 proteins (with minimum one unique or razor peptide).

Principal component analysis (PCA) was performed and the protein expressions for samples were plotted (Figure 1). PCA shows clustering of samples according to their time points with a few exceptions. Overall, the clusters are separated on the principal component 1 (PC1), which has captured the highest proportion of variance (32.3%) in the dataset. Clusters that are formed by the samples at time points 0 hours and 312 hours, shown in Figure 1 indicated by suffixes H0 and H312 respectively, locate close together suggesting a return to pre-infection levels. The samples at time point 81 hours are the farthest away from the samples at time point 0 hours suggesting a higher point of inflammation. Samples from cow C3 suggest a delayed response that correlated with cytokine profiles and clinical observations, while samples from cow C5 suggest an earlier response to the infection.

ANOVA testing was performed to find the proteins that are differentially expressed by a fold change greater than 2 combined with a false discovery rate (FDR)-adjusted P-value less than 0.05 between the 0 hours and the rest of the time points. The highest number of differentially expressed proteins (125 in number) were in 0 hours vs 81 hours time points, while 0 hours vs 36 hours, 0 hours vs 42 hours, 0 hours vs 57 hours and 0 hours vs 312 hours time points showed 20, 43, 66, and 25 proteins respectively. Selected proteins that were up-regulated compared to 0 hours are: serum amyloid A, cathelicidin, haptoglobin, lipopolysaccharide-binding protein, serpin, alpha-enolase, adenylyl

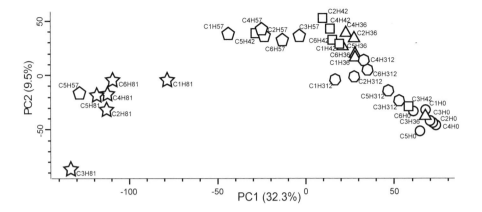

Figure 1. Principal component analysis plots of the proteomics data generated from all 36 milk samples in the challenge study. PC1 and PC2 are plotted in X and Y axes respectively, and captures 32.3 and 9.5% of total variance respectively in the dataset. The labels for each data points refer to the cow from which the milk sample was collected and the time point of the sample collection. The different shapes of the data points show the time point group of the samples.

cyclase-associated protein and vimentin. Only a few proteins were down-regulated, including alpha lactalbumin.

LC-MS based quantitative untargeted metabolomics

Combining the metabolomics data from both the positive and the negative ionisation modes, Ideom software determined 3,828 peaks in the dataset, and from that 740 metabolites were identified and quantified. PCA was again performed and the metabolite expressions for samples were plotted. Similar to the results obtained for the proteomics dataset, the PCA showed clustering of samples according to their time points with a few samples clustering with different time points. The cluster formed by all the samples at time point 0 hours was distinctly separated from clusters formed by samples at all other time points. As suggested by the proteomics data, metabolomic analysis also showed a delayed response by cow C3 and an earlier response by cow C5.

Conclusions

The study shows high concordance between proteomics and metabolomics datasets at global expression levels. Temporal omics data profiling from individual cows correlate well with the early or late clinical response to the infection. The study identified many proteins and metabolites that change during the time course in response to the infection.

References

Boehmer, J. L. 2011. Proteomic analyses of host and pathogen responses during bovine mastitis. *J Mammary Gland Biol Neoplasia* 16 (4): 323-38. http://dx.doi.org/10.1007/s10911-011-9229-x.

Cox, J., and M. Mann. 2008. MaxQuant enables high peptide identification rates, individualized p.p.b.-range mass accuracies and proteome-wide protein quantification. *Nat Biotechnol* 26 (12): 1367-72. http://dx.doi.org/10.1038/nbt.1511.

Creek, D. J., A. Jankevics, K. E. Burgess, R. Breitling, and M. P. Barrett. 2012. IDEOM: an Excel interface for analysis of LC-MS-based metabolomics data. *Bioinformatics* 28 (7): 1048-9. http://dx.doi.org/10.1093/bioinformatics/bts069.

Eckersall, P. D., A. M. de Almeida, and I. Miller. 2012. Proteomics, a new tool for farm animal science. *J Proteomics* 75 (14): 4187-9. http://dx.doi.org/10.1016/j.jprot.2012.05.014.

Tassi, R., T. N. McNeilly, J. L. Fitzpatrick, M. C. Fontaine, D. Reddick, C. Ramage, M. Lutton, Y. H. Schukken, and R. N. Zadoks. 2013. Strain-specific pathogenicity of putative host-adapted and nonadapted strains of Streptococcus uberis in dairy cattle. *J Dairy Sci* 96 (8): 5129-45. http://dx.doi.org/10.3168/jds.2013-6741.

Proteomic profiling of the obligate intracellular bacterial pathogen *Ehrlichia ruminantium* outer membrane fraction

Amal Moumène[1,2,3], Isabel Marcelino[4,5], Miguel Ventosa[4,5], Olivier Gros[3], Thierry Lefrançois[1,2], Nathalie Vachiéry[1,2], Damien F. Meyer[1,2*] and Ana V. Coelho[5*]

[1]CIRAD, UMR CMAEE, F-97170 Petit-Bourg, Guadeloupe, France; damien.meyer@cirad.fr
[2]INRA, UMR1309 CMAEE, F-34398, Montpellier, France
[3]Université des Antilles et de la Guyane, 97159 Pointe-à-Pitre cedex, Guadeloupe, France
[4]IBET, Apartado 12, 2780-901 Oeiras, Portugal
[5]Instituto de Tecnologia Química e Biológica António Xavier, Universidade Nova de Lisboa, Av. da República, 2780-157 Oeiras, Portugal; varela@itqb.unl.pt

Ehrlichia ruminantium is the Gram-negative obligate intracellular bacterium that causes heartwater, a fatal tick-borne disease of ruminants, which is found in the islands of the Indian Ocean and the Caribbean, and in Africa, where it poses important economical constraints to livestock production. A preliminary whole proteome analysis of elementary bodies, the extracellular infectious form of the bacterium, had been performed previously (Marcelino *et al.*, 2012), however due to experimental constrains the detection of outer membrane proteins (OMPs) was limited. These proteins play a crucial role in virulence and pathogenesis of bacteria, and their identification is crucial for vaccine development. Identification of OMPs is also essential for understanding Ehrlichia's OM architecture, and how the bacterium interacts with the host cell environment. To gain insight into the protein composition of the ehrlichial OM, we performed the proteome characterization of OM fractions from *E. ruminantium* elementary bodies. We started by developing a new enrichment protocol for ER outer membranes using the ionic detergent sarkosyl, followed by 1DE separation of the solubilized protein extract. The proteome profiling was done analyzing tryptic peptide mixtures by LC-MALDI-TOF/TOF.

Of 46 unique proteins identified in the OM fraction, 18 (39%) were OMPs, including 8 proteins involved in cell structure and biogenesis, 4 in transport/virulence, 1 porin, and 5 proteins of unknown function. This work represents the most complete proteome characterization of the OM fraction in Ehrlichia spp. The results obtained indicate that the subcellular fractionation developed is suitable for OM proteome profiling. Some of the identified proteins are potentially involved in *E. ruminantium* pathogenesis, which are good novel targets for candidate vaccines. In summary, we provide both pioneering data and novel insights into the pathogenesis of this obligate intracellular bacterium.

References

Marcelino, I., A Martinho de Almeida, C Brito, DF Meyer, M Barreto, C Sheikboudou, CF Franco, D Martinez, T Lefrançois, N Vachiéry, MJT Carrondo, AV Coelho, PM Alves, 2012. Proteomic analyses of *Ehrlichia ruminantium* highlight diferential expression of MAP1-family proteins in host endothelial cells. Veterinary Microbiology 156: 305-14

NMR for metabolomic tissue profiling in small ruminants: a tool for proteomics study complementation

Mariana Palma[1]*, Manolis Matzapetakis[1] and André M. Almeida[2,3,4]

[1]Instituto de Tecnologia Química e Biológica (I.T.Q.B.), Universidade Nova de Lisboa. Av. da República, Estação Agronómica Nacional, 2780-157 Oeiras, Portugal; mpalma@itqb.unl.pt

[2]Instituto de Investigação Científica Tropical (I.I.C.T.), Rua da Junqueira, no. 86 – 1, 1300-344 Lisboa, Portugal

[3]Instituto de Biologia Experimental Tecnológica (I.B.E.T.). Av. da República, Estação Agronómica Nacional, 2780-157 Oeiras, Portugal

[4]Centro Interdisciplinar de Investigação em Sanidade Animal (C.I.I.S.A), Faculdade de Medicina Veterinária, Universidade de Lisboa, Avenida da Universidade Técnica 1300-477 Lisboa, Portugal

Introduction

Goat and sheep's meat, milk and dairy products are a valuable nutritional supply, especially in developing countries, where they are important sources of animal protein (Blümmel *et al.*, 2010).

Understanding the physiology of small ruminants is of utmost interest as this information could be used for genetic improvement towards the increase of production yields.

Metabolomics studies provide an overview of the small metabolites present and their concentration in the sample, allowing a more general analysis of the state of the tissue/animal. Therefore, metabolomics could be used as a complementary tool to help in the proteomics analysis.

The development of new methodologies that allow accessing metabolome information could be very useful in this systemic approach. We are applying Nuclear Magnetic Resonance (NMR) techniques to study tissue metabolomes, using samples from small ruminants of different origin, to test the efficiency of the protocol and the possibility of returning useful data in this field.

We tested various protocols on muscle samples from sheep and mammary gland samples from goat, to identify the optimal methodologies for each tissue and investigate the robustness of the results after statistical analysis.

Material and methods

For this work we used sets of eight sheep (*Ovis aries*) muscle samples from each of the Damara and Merino breeds (Alves *et al.*, 2013); as well as mammary gland samples from goat (*Capra hircus*): six from the Palmera and four from the Majorera breed (Lérias *et al.*, 2013). Sheep were slaughtered and the gastrocnemius muscle was sampled and preserved at -80 °C until further use.

Mammary gland samples were collected by biopsy and preserved at -80 °C. All samples were treated individually in all the procedures herein described. Frozen samples were ground to powder using mortar and pestle. The mass of the samples varied between 162.2 mg and 833.1 mg in muscle and 43.3 mg and 140.0 mg in the mammary gland. Metabolite extraction was performed using the Bligh and Dyer method (Bligh and Dyer, 1959) with some adaptations, to separate its aqueous and the organic fraction. The organic fraction was dried under a nitrogen flow and kept at -80 °C, and the aqueous fraction was dried in a vacuum evaporator and preserved at -80 °C until preparation for NMR analysis. In this work, only aqueous fractions were analysed.

Samples were prepared by res-suspending the previously dried aqueous fractions in 600 µl of 150 mM phosphate buffer pH 7.0, containing 100 µM DSS and 10% deuterated water. After centrifugation for the precipitation of small particulates samples were transferred to 5 mm NMR tubes.

All NMR spectra were acquired on a 800 MHz Bruker AVANCE II$^+$ spectrometer equipped with a room temperature triple resonance HCN Z-gradient probe, at 298 K. For each sample an ^{1}H 1D-NOESY and a CPMG spectrum were collected while for the most concentrated sample of each series an additional set of spectra to assist with the resonance was collected overnight consisting of a J-Resolved, ^{1}H-COSY and an ^{1}H-^{13}C HSQC. Each sample required less than 2 hours of data acquisition making this approach an easy and quick way to access the metabolome information.

Spectra were processed and analysed using TopSpin 2.1. The variability within as well as among the sets of spectra were initially evaluated using binning and Principal Component Analysis (PCA), with the R Software. Identification and quantification (profiling) of the sample components was performed with the Chenomx NMR Suite 7.1 Software and the 'BATMAN' and 'chemometrics' packages of 'R'.

Results and discussion

The protocols for extraction and data acquisition were shown to be robust and reproducible as seen from repetitions and by observing the ratio of selected compound concentrations as a function of tissue mass. We also performed spiking for verification which also confirmed that the level of extraction was very reproducible.

Statistical analysis showed that tissue extracts from muscle and mammary gland can be easily distinguished and are clearly grouped by PCA analysis in two different clusters. These results confirm, not only that the composition of each metabolome is different, but also that the applied procedures are adequate to this type of analysis. Furthermore, this methodology seems to be adequate for profiling, allowing the identifications of around thirty compounds in each tissue as seen in Figure 1.

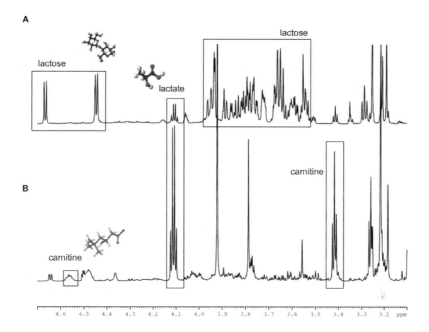

Figure 1. ^1H 1D NOESY Spectra (central region) from aqueous fraction of goat mammary gland (A) and sheep muscle (B), with examples of identified compounds. (molecule images from Biological Magnetic Resonance Data Bank).

This work showed that NMR metabolomics, which is a relatively novel technique in animal science, represents a promising methodology to assess metabolome information. It is also able to evaluate differences between groups of samples which can effectively complementing proteomic analysis and reveal if differences in the proteome translate into the metabolome.

References

Alves, S.P., Bessa, R.J.B., Quaresma, M.A.G., Kilminster, T., Scanlon, T., Oldham, C., Milton, J., Greeff, J. and Almeida, A.M., 2013. Does the fat tailed Damara ovine breed have a distinct lipid metabolism leading to a high concentration of branched chain fatty acids in tissues? PLOS ONE 8(10), e77313.

Bligh, E.G. and Dyer, W.J., 1959. A rapid method of total lipid extraction and purification. Canadian Journal of Biochemistry and Physiology 37, 911-917.

Blümmel, M., Wright, I.A. and Hegde, N.G., 2010. Climate change impacts on livestock production and adaptation strategies: a global scenario. National Symposium on Climate Change and Rainfed Agriculture, Hyderabad, India, February 18-20.

Lérias, J.R., Hernández-Castellano L.E., Morales-delaNuez A., Araújo S.S., Castro N., Argüello, A., Capote, J. and Almeida, A.M., 2013. Body live weight and milk production parameters in the Majorera and Palmera goat breeds from the Canary Islands: influence of weight loss. Tropical Animal Health and Production 45, 1731-1736.

Part II
Proteomics in infection diseases and animal production

Proteomics for assessing quality and metabolic compatibility of aquaculture feeds

M. Filippa Addis[1*], Roberto Anedda[1], Grazia Biosa[1], Elia Bonaglini[1], Roberto Cappuccinelli[1], Stefania Ghisaura[1], Riccardo Melis[1], Daniela Pagnozzi[1], Simona Spada[1], Hanno Slawski[2] and Sergio Uzzau[1]

[1]Porto Conte Ricerche, SP 55 Porto Conte/Capo Caccia, Loc. Tramariglio, 07041 Alghero, Sassari, Italy; addis@portocontericerche.it
[2]Aller Aqua, Allervej 130, 6070 Christiansfeld, Denmark

Introduction

Aquaculture contributes to about half of the fish consumed worldwide, and it is currently witnessing a growth higher than any other animal food-producing sector. In order to contain feed costs and to contrast the unsustainable overexploitation of the wild fish stocks for feed production, there is a need to develop new formulations incorporating other nutrients, such as vegetable matter, other economically valid integrations, or nutrients obtained as byproducts from other food production industries. In any case, this substitution should still ensure a good compatibility with fish metabolism in order to maintain a satisfactory production efficiency (conversion factor) and product quality.

The characterization of differential protein expression profiles in farmed fish tissues can help gain a better understanding of growth dynamics and compatibility of feeds with fish metabolism, and also present valuable implications for product quality, food safety and authentication, traceability, and shelf-life, facilitating monitoring, control, and optimization of aquaculture practices (Addis *et al.*, 2010, 2012; Addis, 2013; Campus *et al.*, 2010; Ghisaura *et al.*, 2014; Terova *et al.*, 2011). In our laboratories, an integrated approach combining biometrics, lipidomics, metabolomics and proteomics was implemented with the aim of assessing the impact of different feeds on the zootechnical performance and metabolism of gilthead sea breams (*Sparus aurata*, L.) (Anedda *et al.*, 2013; Melis *et al.*, 2014a,b), one of the most valuable products of Mediterranean aquaculture.

Materials and methods

Fish farming was carried out in controlled conditions in the Blue Biotechnology Labs at Porto Conte Ricerche, using three independent seawater circuits each composed of three 550 L tanks. Trials lasted about 12 weeks for each feed. Commercial feeds lacking a satisfactory level of information on formulation in their labels were characterized by shotgun proteomics by FASP and LTQ-Orbitrap. At completion, zootechnical performances were assessed by morphometry, and fish muscle was subjected to proximate analysis and lipidomic/metabolomic fingerprinting by GC-MS and NMR. Liver and blood serum protein expression profiles were investigated by 2D DIGE and DeCyder analysis, LC-MS/MS for protein identification, and Ingenuity Pathway Analysis.

Results and discussion

Up to now, the complete approach has been applied to several feeding trials. In the first round, three differently substituted feed formulations available in the market were evaluated upon request of the Sardinian Fish Farmers' Association (Associazione Acquacoltori Sardi). Characterization of feeds by shotgun proteomics and label-free quantitation provided useful data on their composition in terms of vegetable and animal proteins, and analysis by GC-MS produced detailed information on their lipid composition. Upon 2D DIGE/MS of fish liver and blood serum, feed-specific changes in protein expression were identified, especially related to the lipid and aminoacid metabolisms, that helped the identification of the best performing feed. In fact, hints on their compatibility with fish metabolism were obtained, as well as indications on potential problems deriving from the lack of certain nutrients (essential aminoacids and lipids) or the excess of others (starch), such as the increase in amino acid metabolism enzymes, oxidative stress pathways, and alterations in the small molecule metabolism. In the second round, based on findings emerged from the first one, the supplementation with methionine of one of the feeds from the first trial was assessed. In this case, we found that such integration was able to contrast some alterations in protein expression patterns that were observed when the non-integrated feed was used. According to Ingenuity Pathway Analysis, protein synthesis, amino acid metabolism, and small molecule biochemistry were the metabolic networks most affected by methionine supplementation.

The proteomic analysis pipeline is now being applied in our research centre in support of Aller Aqua, a producer of high quality feeds for aquaculture. In a recently completed study, three different feeds were assessed: vegetable base with fish flour, vegetable base with fish flour and meat meal, and vegetable base with fish flour and echinoderm flour. As a result, the three feeds gave rise to statistically significant differences in the liver protein profiles. In general, the liver enzymes undergoing the most significant alterations had similar identities and trends to those previously observed, and highlighted a better performance of the echinoderm-integrated feed in terms of metabolic compatibility. Currently, the performance of feeds optimized for colder and warmer water temperatures is being evaluated in a more elaborate feeding trial involving changes in both the feeds and the water temperature. The feeds being evaluated are specifically designed for use in the different seasonal atmospheric conditions.

Currently, a shotgun proteomics approach employing FASP and LTQ-Orbitrap Velos MS is also being applied to fish muscle, and its potential for studying the impact of fish feeds on muscle protein composition along fish growth is being assessed. In this case, the application of label-free proteomics will need to be carefully evaluated when considering that the protein sequence databases for many farmed fish species are not complete.

Conclusion

Proteomics can represent a valid support to biometrics, lipidomics and metabolomics when investigating feed compatibility with the farmed fish metabolism. In our laboratories, the proteomic study of liver and serum protein profiles is providing useful insights on the changes occurring in the lipid, carbohydrate, amino acid and small molecule pathways, as well as on the impact on oxidative stress, when fish are fed different formulations. Specifically, liver proteomics can highlight changes occurring in metabolic pathways, possibly providing indications on how to improve feed quality and composition in terms of compatibility with the farmed fish species. Serum proteomics, although requiring further optimization and investigation, might hold promise as a tool for rapidly monitoring metabolic changes along farming, with opportunities for correcting the feeding regimen adopted. In any case, reaching the optimal balance between feed price, weight gain, and product quality is of course pivotal for maximizing the economic value of a farming plant. Our final aim is to implement proteomics in a multidisciplinary approach aimed to support the aquaculture industry in maintaining the best balance between production efficiency and product quality.

References

Addis, M.F. 2013. Farmed and wild fish. In: Toldrá, F. and Nollet, L.M.L. (ed.) Proteomics in Foods: Principles and Applications. Springer, New York, USA, pp 181-203.

Addis, M.F., Cappuccinelli, P., Tedde, V., Pagnozzi, D., Porcu, M.C., Bonaglini, E., Roggio, T., and Uzzau S. 2010. Proteomic analysis of muscle tissue from gilthead sea bream (*Sparus aurata*, L.) farmed in offshore floating cages. Aquaculture 309, 245-252.

Addis, M.F., Pisanu, S., Preziosa, E., Bernardini, G., Pagnozzi, D., Roggio, T., Uzzau, S., Saroglia, M., and Terova, G. 2012. 2D DIGE and MS for investigating the influence of slaughtering on post-mortem integrity of fish filet proteins. Journal of Proteomics 75, 3654-64.

Anedda, R., Piga, C., Santercole, V., Spada, S., Bonaglini, E., Cappuccinelli, R., Mulas, G., Roggio, T., and Uzzau S. 2013. Multidisciplinary analytical investigation of phospholipids and triglycerides in offshore farmed gilthead sea bream (*Sparus aurata*) fed commercial diets. Food Chemistry 138, 1135-1144.

Campus, M., Addis, M.F., Cappuccinelli, R., Porcu, M.C., Pretti, L., Tedde, V., Secchi, N., Stara, G., and Roggio, T. 2010. Stress relaxation behaviour and structural changes of muscle tissues from gilthead sea bream (*Sparus aurata* L.) following high pressure treatment. Journal of Food Engineering 96, 192-198.

Ghisaura, S., Anedda, R., Pagnozzi, D., Biosa, G., Spada, S., Bonaglini, E., Cappuccinelli, R., Roggio, T., Uzzau, S., and Addis M.F. 2014. Impact of three commercial feed formulations on farmed gilthead sea bream (*Sparus aurata*, L.) metabolism as inferred from liver and blood serum proteomics. Proteome Science 12, 44.

Melis, R., and Anedda, R. 2014. Biometric and metabolic profiles associated to different rearing conditions in offshore farmed gilthead sea bream (*Sparus aurata* L.). Electrophoresis 35, 1590-1598.

Melis, R., Cappuccinelli, R., Roggio, T., and Anedda, R. 2014. Addressing marketplace gilthead sea bream (*Sparus aurata* L.) differentiation by ^{1}H NMR-based lipid fingerprinting. Food Research International 63, 258-264.

Terova, G., Addis, M.F., Preziosa, E., Pisanu, S., Biosa, G., Pagnozzi, D., Gornati, R., Bernardini, G., Roggio, T., and Saroglia, M. 2011. Effects of postmortem storage temperature on sea bass (*Dicentrarchus labrax*) muscle protein degradation: analysis by 2D DIGE and mass spectrometry. Proteomics 11, 2901-2910.

Finding novel milk protein markers for small ruminant mastitis

Maria Filippa Addis[1]*, Salvatore Pisanu[1], Vittorio Tedde[1], Stefania Ghisaura[1], Grazia Biosa[1], Daniela Pagnozzi[1], Tiziana Cubeddu[2], Stefano Rocca[2], Gavino Marogna[3], Ignazio Ibba[4], Marino Contu[4], Simone Dore[5], Agnese Cannas[5] and Sergio Uzzau[1]

[1]Porto Conte Ricerche, SP 55 Porto Conte/Capo Caccia, Loc. Tramariglio, 07041 Alghero, Italy; addis@portocontericerche.it

[2]Department of Veterinary Medicine, University of Sassari, Via Vienna 2, 07100 Sassari, Italy

[3]Istituto Zooprofilattico Sperimentale della Sardegna 'G. Pegreffi', Via Duca degli Abruzzi 8, 07100 Sassari, Italy

[4]Associazione Regionale Allevatori della Sardegna, Via Cavalcanti 8, 09128 Cagliari, Italy

[5]Centro di Riferimento Nazionale per le Mastopatie degli Ovini e dei Caprini, Via Duca degli Abruzzi 8, 07100 Sassari, Italy

Introduction

Small ruminant farming for dairy production plays a major role in the economy of several Mediterranean areas, and forms the basis for numerous valuable traditional dairy products. The milk somatic cell count (SCC) for monitoring udder health and milk quality, widely used and established in cattle, is being applied also to small ruminants, but the question of which threshold levels are appropriate in these animals is still lively and strongly debated. In fact, adding to intramammary infections, other factors impact on SCCs in both sheep and goats, including breed, parity, stage of lactation, milking techniques, management practices, grazing style, environmental factors, feeding regimens, and others, often with significant changes and with wider ranges than those observed in cows. In addition, the smaller dimensions of the lactating organ, the shorter lactation time and the reduction in milk volume at the end of lactation, as well as the higher incidence of infections by gram-positive bacteria, further complicate the picture (Bergonier *et al.*, 2003; Contreras *et al.*, 2007; Marogna *et al.*, 2010).

In this context, it appears evident that, for a thorough and sensitive monitoring of udder health, as well as for detecting the more insidious subclinical mastitis cases, SCCs would benefit of the integration with additional markers, also for validation purposes and for a better definition of diagnostic thresholds. In our laboratories, and with the collaboration of the local animal health institution (Istituto Zooprofilattico della Sardegna, IZS) and the Department of Veterinary Sciences at the University of Sassari, a combined gel-based proteomic study was carried out on milk and mammary tissues from sheep suffering natural and experimental mammary infections by two of the most relevant bacterial agents of sheep mastitis, *Streptococcus uberis* and *Mycoplasma agalactiae*. The main aim of our study was to identify novel proteins or peptides able to support monitoring of milk quality and udder health in sheep, with potential application also in dairy goats. Moreover, the identification of proteins or peptides released in milk specifically upon infection will hopefully contribute to collect further evidence on the significance of the SCC alterations in sheep and goat milk, and to elucidate the innate immune processes taking place in the mammary gland of small ruminants. A large scale validation is currently ongoing in collaboration with the Sardinian Farmers' Association (Associazione Regionale Allevatori della Sardegna, ARAS) and the National Reference Center for Sheep and Goat Mastitis (Centro di Referenza Nazionale per le Mastopatie degli Ovini e dei Caprini, CReNMOC).

Materials and methods

Milk from healthy sheep, from sheep naturally infected by *M. agalactiae*, and from sheep experimentally infected with *S. uberis* was subjected to 2D-DIGE-MS followed by ESI-Q-TOF-MS/MS, and by SDS-PAGE followed by band cutting, digestion, and LTQ-Orbitrap Velos, and then by label-free quantitation and Ingenuity Pathway Analysis for characterization of differential proteins (Addis *et al.*, 2011, 2013). The biomarker discovery study was carried out on proteins extracted from the milk fat fraction prepared as described before (Addis *et al.*, 2011; Pisanu *et al.*, 2011, 2012, 2013). The cellular origin of the markers of interest was then investigated by immunohistochemistry (IHC) on infected mammary tissues (Addis *et al.*, 2013). Validation of the putative mastitis biomarkers by western immunoblotting is currently ongoing on a large cohort of sheep and goat milk samples from problematic farms.

Results

The fat layer fraction of mastitic sheep milk showed significantly increased levels of proteins and peptides involved in inflammation and immune defense (such as antimicrobial proteins and peptides), in folding (such as HSPs), and in vesicular trafficking, when compared to milk from healthy sheep. Increased amounts of proteins involved in oxidative stress were also observed. On the other hand, typical milk fat globule and milk proteins, including butyrophilin, lactadherin, adipophilin, and xanthine dehydrogenase/oxidase, as well as alpha-S1-casein, beta-casein, and kappa-casein, were all present in significantly lower amounts.

Further investigations on mastitis biomarkers focused on secreted antimicrobial and immune defense proteins, since these are released in milk, are likely secreted in a non-strictly pathogen-dependent fashion, are produced since the first stages of infection, and are inflammation-specific. Among these, the most significant changes were seen for lactotransferrin, cathelicidins, the calprotectin subunits S100A8 and S100A9, myeloperoxidase, complement C3, haptoglobin/zonulin, immune cell-related proteases, pentraxin-related protein PTX3, neutrophil cytosolic factors, serotransferrin, and bactericidal permeability-increasing protein. Aiming to elucidate the innate immune mechanisms at play in the infected mammary gland, especially concerning the cellular sources and production kinetics of antimicrobial and immune defense proteins, IHC assays were carried out for a few selected proteins on mammary tissues with patterns resembling acute infections and subclinical infections. In the first case, strongly positive signals for the biomarkers investigated were present in immune cells filling the alveolus lumen and in their surroundings. In the second case, positivity was seen in lactocytes lining the milk ducts, indicating the active involvement of epithelial cells in the first stages of innate immune response. Interesting insights on formation of neutrophil extracellular traps in sheep milk and how their components can be exploited as mastitis biomarkers were also obtained upon IHC and proteomic data analysis. A large scale validation study of a selected panel of potential mastitis markers is currently ongoing on hundreds of samples from sheep flocks affected by bacterial mastitis of different etiologies. As a preliminary result, positivity of the markers under evaluation is steadily associated with high SCCs, while the opposite is not always true. In addition, the correlation between markers and SCC levels is starting to become apparent. Last, but not least, the biomarkers under evaluation in sheep are showing similar patterns also in goats.

Conclusions

The biomarker discovery approach applied to milk from sheep suffering mastitis from different etiologic agents enabled the identification of several proteins involved in the innate immune response that are released in a rapid and specific fashion in milk upon infection. These molecules hold promise as novel sheep mastitis markers, and open valuable perspectives for the development of diagnostic tools enabling a better monitoring of udder health and milk quality in sheep and goat farms. In addition, these novel markers have potential for cross-validation or integration of the already used SCC parameter in small ruminants.

References

Addis, M.F., Pisanu, S., Ghisaura, S., Pagnozzi, D., Marogna, G., Tanca, A., Biosa, G., Cacciotto, C., Alberti, A., Pittau, M., Roggio, T., and Uzzau, S. 2011. Proteomics and pathway analysis of the milk fat globule in sheep naturally infected by Mycoplasma agalactiae: insights into the *in vivo* response of the mammary epithelium to bacterial infection. Infection and Immunity 79, 3833-3845.

Addis, M.F., Pisanu, S., Marogna, G., Cubeddu, T., Pagnozzi, D., Cacciotto, C., Campesi, F., Schianchi, G., Rocca, S., and Uzzau, S. 2013. Production and release of immune defense proteins by mammary epithelial cells following Streptococcus uberis infection of sheep. Infection and Immunity 81, 3182-3197.

Bergonier, D., De Crémoux, R., Rupp, R., Lagriffoul, G., and Berthelot, X. 2003. Mastitis of dairy small ruminants, Veterinary Research 34, 689-716.

Contreras, A., Sierra, D., Sánchez, A., Corrales, J.C., Marco, J.C., Paape, M.J., and Gonzalo, C. 2007. Mastitis in small ruminants. Small Ruminant Research 68, 145-153.

Marogna, G., Rolesu, S., Lollai, S., Tola, S., and Leori, G. 2010. Clinical findings in sheep farms affected by recurrent bacterial mastitis. Small Ruminant Research 88, 119-125.

Pisanu, S., Ghisaura, S., Pagnozzi, D., Falchi G., Biosa, G., Tanca, A., Roggio, T., Uzzau, S., and Addis, M.F. 2012. Characterization of sheep milk fat globule proteins by two-dimensional polyacrylamide gel electrophoresis/ mass spectrometry and generation of a reference map. International Dairy Journal 24, 78-86.

Pisanu, S., Ghisaura, S., Pagnozzi, D., Tanca, A., Biosa, G., Roggio, T., Uzzau, S., and Addis, M.F. 2011. The sheep milk fat globule membrane proteome. Journal of Proteomics 74, 350-358.

Pisanu, S., Marogna, G., Pagnozzi, D., Piccinini, M., Leo, G., Tanca, A., Roggio A.M., Roggio, T., Uzzau, S., and Addis, M.F. 2013. Characterization of size and composition of milk fat globules from Sarda and Saanen dairy goats. Small Ruminant Research 109, 141-151.

Proteomic identification of immunogenic proteins in *Leishmania infantum*

Katarina Bhide[1,2], Carmen Aguilar-Jurado[1], Sara Zaldivar-Lopez[1], Ignacio López-Villalba[4], Manuel Sánchez-Moreno[5], Ángela Moreno[1,3] and Juan J. Garrido[1*]

[1]*Animal Breeding and Genomics Group, Department of Genetics, University of Córdoba, 14071 Córdoba, Spain; ge1gapaj@uco.es*

[2]*Department of Microbiology and Immunology, University of Veterinary Medicine and Pharmacy, Košice, Slovakia*

[3]*Institute for Sustainable Agriculture, CSIC, P.O. Box 4084, 14004, Córdoba, Spain*

[4]*Department of Animal Medicine and Surgery, University of Córdoba, 14071 Córdoba, Spain*

[5]*Department of Parasitology, University of Granada, 18071 Granada, Spain*

Introduction

In the Mediterranean and Middle East, canine leishmaniasis is an important emerging zoonosis, caused by *Leishmania infantum*, a vector-borne protozoan parasite. The parasite is generally transmitted by the bite of the phlebotomine sandfly and occurs in two forms: the intracellular amastigote form found in the host, and the promastigote form predominately found in the insect vector (Rosypal *et al.*, 2005). Although there is a minimal risk of transmission through other routes, transfusion-mediated and transplacental transmissions have been also reported (Boggiatto *et al.*, 2011).There are two clinical forms of the disease: cutaneous and visceral. Clinical signs of the cutaneous form are alopecia, skin lesions and ulcerative or exfoliative dermatitis. Visceral form is characterized by epistaxis, renal and hepatic failure, progressive loss of weight with decreased appetite and swollen lymph nodes. Dogs are the main parasite reservoir although leishmaniasis also affects humans and other domestic animals. In the present work, an immunoproteomic approach, together with two-dimensional electrophoresis (2DE) and mass spectrometry, was carried out to identify antigenic proteins recognized by antibodies present in the sera of dogs with symptomatic leishmaniasis. The identified proteins constitute a significant source of information for the improvement of diagnostic tools and/or vaccine development to the disease.

Materials and methods

Leishmania promastigotes were grown in BHI medium at 26 °C for one week. Cells were harvested and proteins were extracted with the help of a lysis buffer (7M urea, 2M thiourea, 4% CHAPS, 30mM DTT, 1% PMFS, 50 U DNAsc I per ml of lysis buffer, proteinase inhibitor cocktail, 1% ampholytes) and separated by 2DE. The concentration of proteins in supernatant was measured by Bradford assay by making known-serial dilution of BSA (bovine serum albumin) as the standard protein (BioRad). Proteins were separated by isoelectric focusing in precast IPG strips (17 cm, pH 4-7, BioRad). Conditions for focusing were rehydration 20 °C/12 h, 250 V linear 15 min, 500 V linear 30 min, 1000 V linear 1 h, 2000 V linear 1 h, 5000 V linear 2 h, 8000 V linear 1 h 30 min and 80

000 V rapid volts/hours. After idoacetamide and DTT treatment of proteins in strips, the second dimension was performed on 12% acrylamide gels in running buffer at 60 mA for 45 minutes and then 120 mA 4 hours. One of the gels was stained using colloidal Coomassie Brilliant Blue G-250 dye according to manufacturer's instructions (BioRad) and the second gel was used for protein transfer to PVDF transfer membrane (PVDF Immobilon-P Membrane, 0.45 μm, Millipore) in transfer buffer at 30 volts at 4 °C overnight. After transfer, non-specific binding sites on membrane were blocked with 1% skimmed milk in TTBS (25 mM TRIS, 150 mMNaCl, 0.05% Tween 20, pH 7.2) for 1 hr. Membrane was then incubated with dog serum (1:1000 in 1% skim milk in TTBS) pooled from 5 dogs naturally infected with *Leishmania infantum*, all of them with advanced clinical disease. After three washings, membranes were incubated with goat anti-dog antibody conjugated with HRP diluted 1:50000 in 1% skim milk in TTBS. Antigen-antibody complex was visualized with ECL using Amersham prime western blotting detection reagent (GE Healthcare). Signals were captured on LAS-3000 imaging system (Fujifilm). Protein spots observed on membrane were correlated/aligned with the protein spots on the gel stained with Coomassie blue. Spots were manually picked and transferred to tubes with 500 μl of MilliQ water. Protein spots were identified by peptide mass fingerprinting at the Proteomic Service of the University of Córdoba.

Results and discussion

The present study applied an immunoproteomic approach in *Leishmania infantum* promastigote-like antigenic extracts, using pools of sera of infected dogs, in an attempt to identify new targets as diagnosis and/or vaccine candidates.

The analysis of 38 protein spots (Figure 1) allowed the identification of 33 well-defined proteins (Table 1) that were recognized by the sera of symptomatic dogs for leishmaniasis. Among these

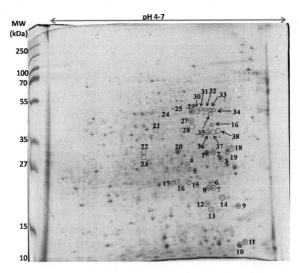

Figure 1. 2D gel of protein extracted from Leishmania infantum. *Circled protein spots were picked and subjected to MALDI-TOF-MS for identification.*

Table 1. Proteins of Leishmania infantum *promastigotes identified using an immunoproteomic approach.*

Spot no.	Protein score	Protein name	Species	Protein Mw / PI	Accession no.
1	652	beta tubulin	*Leishmania infantum* JPCM5	50.3479 kDa / 4.74	gi|146086185
2	293	beta tubulin	*Leishmania infantum* JPCM5	50.3479 kDa / 4.74	gi|146086185
3	273	beta tubulin	*Leishmania infantum* JPCM5	50.3479 kDa / 4.74	gi|146086185
4	213	Eukaryotic translation initiation factor 6 (eIF-6)	*Leishmania braziliensis* MHOM/BR/75/M2904	27.4089 kDa / 5.05	gi|154345574
5	347	Eukaryotic translation initiation factor 6 (eIF-6)	*Leishmania braziliensis* MHOM/BR/75/M2904	27.4089 kDa / 5.05	gi|154345574
6	213	activated protein kinase C receptor homolog LACK	*Leishmania donovani*	35.085 kDa / 5.54	gi|3132790
10	331	calpain-like cysteine peptidase	*Leishmania braziliensis* MHOM/BR/75/M2904	13.1613 kDa / 4.7	gi|154334241
11	176	beta tubulin	*Leishmania major*	50.3379 kDa / 4.75	gi|1296834
14	137	eukaryotic initiation factor 5a	*Leishmania braziliensis* MHOM/BR/75/M2904	17.9288 kDa / 4.83	gi|154338682
16	94	hypothetical protein	*Leishmania infantum* JPCM5	30.7232 kDa / 5.8	gi|146089119
18	90	translation elongation factor 1-beta	*Leishmania braziliensis* MHOM/BR/75/M2904	22.9675 kDa / 4.56	gi|154345682
19	135	chaperonin 60.2 precursor	*Leishmania donovani*	60.8705 kDa / 5.32	gi|4323557
20	831	alpha tubulin	*Leishmania braziliensis* MHOM/BR/75/M2904	54.2617 kDa / 5.15	gi|154333816
24	91	trypanothione reductase	*Crithidia fasciculata*	43.0736 kDa / 5.29	gi|552306
26	280	heat shock protein	*Leishmania braziliensis*	80.3644 kDa / 5.03	gi|62944644
27	632	heat shock protein	*Leishmania braziliensis*	80.3644 kDa / 5.03	gi|62944644
28	184	ATPase beta subunit	*Leishmania infantum* JPCM5	56.485 kDa / 5.14	gi|146088806
29	851	heat shock protein 70	*Leishmania infantum*	72.1612 kDa / 5.35	gi|239580147
30	212	heat shock protein 70	*Leishmania donovani*	45.1741 kDa / 5.36	gi|209166099
31	130	ATPase beta subunit	*Leishmania major* strain Friedlin	56.513 kDa / 5.14	gi|157870646
32	174	glucose-regulated protein 78	*Leishmania braziliensis* MHOM/BR/75/M2904	72.0373 kDa / 5.06	gi|154340409
37	709	ATPase beta subunit	*Leishmania major* strain Friedlin	56.513 kDa / 5.14	gi|157870646

proteins, the presence of elongation factors; HSP70 and tubulin, among the most abundantly detected in the antigenic extracts, were in good agreement with other studies and present a reliable validation of the immunoproteomic analysis performed herein. Several of the identified proteins are linked to metabolism pathways, such as ATP-synthase and heat shock proteins (Coello *et al.*, 2012). Others have been considered therapeutic targets, such as cysteine peptidases and eIF-6 (Poot *et al.*, 2006), and some of them are involved in parasite virulence, such as LACK antigen (Kelly *et al.*, 2003). Obtained results represent a contribution toward the future improvement of diagnostic tools and vaccines for canine leishmaniasis, and also a step towards a better understanding of the biological role of the identified proteins in the pathogenesis of *Leishmania infantum*.

Acknowledgements

Short term scientific mission of KB was supported by a grant from the COST Action on Farm Farm Animal Proteomics (FA1002). KB is supported by APVV-036-10 and ITMS-26220220185.

References

Boggiatto, P.M., Gibson-Corley, K.N., Metz, K., Gallup, J.M., Hostetter, J.M., Mullin, K. and Petersen, C.A., 2011. Transplacental transmission of Leishmania infantum as a means for continued disease incidence in North America. PLoS Negl Trop Dis 5: e1019.

Coelho VTS, Oliveira JS, Valadares DG, Chávez-Fumagalli MA, Duarte MC, *et al.*, 2012. Identification of Proteins in Promastigote and Amastigote-like *Leishmania*Using an Immunoproteomic Approach. PLoS Negl Trop Dis 6(1): e1430.

Kelly BL, Stetson DB, and Locksley RM., 2003. Leishmania major LACK antigen is required for efficient vertebrate parasitization. Journal of Experimental Medicine 198(11): 1689-1698.

Poot J, Spreeuwenberg K, Sanderson SJ, Schijns VECJ, Mottram JC, *et al.*, 2006. Vaccination with a preparation based on recombinant cysteine peptidases and canine IL-12 does not protect dogs from infection with *Leishmania infantum*. Vaccine 24: 2460-2468.

Rosypal, A.C., Troy, G.C., Zajac, A.M., Frank, G. and Lindsay, D.S., 2005. Transplacental transmission of a North American isolate of Leishmania infantum in an experimentally infected beagle. J Parasitol 91: 970-972.

PilE4 of *Francisella* alters expression of proteins on the brain endothelium

Mangesh Bhide[1,2]*, Elena Bencurova[1], Andrej Kovac[2], Lucia Pulzova[1] and Zuzana Flachbartova[1]
[1]*University of veterinary medicine and pharmacy in Košice, Komenskeho 73, Košice, Slovakia;*
bhidemangesh@gmail.com
[2]*Institute of Neuroimmunology of Slovak Academy of Sciences, Dubravska cesta 9, Bratislava, Slovakia*

Objectives

Francisella tularensis is highly infectious Gram-negative facultative intracellular coccobacillus, and causative agent of arthropod-borne zoonotic disease tularemia. It has been detected in over 250 host species, including human. Pathogens exploit several strategies to cross the blood-brain barrier (BBB) i.e. transcellular, paracellular and/or by means of infected phagocytes (Trojan horse mechanism) (Bencurova *et al.*, 2011; Pulzova *et al.*, 2009). Crossing of BBB is a complex process, while bacterial adhesion to brain microvascular endothelial cells (BMECa) is the main prerequisite for initiation of para- or transcellular translocations (Bencurova *et al.*, 2011). Extracellular *Francisella* readily adheres to endothelial and epithelial cells (Lindemann *et al.*, 2007; Melillo *et al.*, 2006; Moreland *et al.*, 2009). Previously we found, that *Francisella* adhere to brain microvascular endothelial cells and this adhesion is mediated by ICAM-1 molecule from the host and PilE4 from the *Francisella* site (Bencurova *et al.*, 2013). To understand the underlying principles of initial adhesion of *Francisella* to BMECs, we used qRT-PCR to investigate the effect of PilE4 to brain endothelium.

Material and methods

Cutlivation of Francisella

F. tularensis subsp. *holarctica* stain LVS was cultivated on chocolate agar with 1% glucose and 0.1% L-cystein at 37 °C for 4 days. Bacteria were harvested by scrubbing, washed with PBS and re-suspended in distilled water. DNA was isolated by heating at 98 °C for 10 minutes and the concentration was measured.

Preparation of recombinant PilE4

Total DNA from LVS was isolated and region encoding PilE4 was amplified using a target specific primers. Obtained amplicon was gel purified and ligated into pQE-30UA-GFP expression vector (His-tag at N-terminal). Vectors were electroporated into *E. coli* SG13009. The cells were grown overnight on the selective LB agar containing ampicillin and kanamycin. The expression of His tag proteins was induced by adding IPTG (1 mM, Sigma-Aldrich, Slovakia). Cells were pelleted by centrifugation and sonicated in the lysis buffer containing 1 mg/ml lysozyme, 1% of nuclease mix and 1% of a protease inhibitor cocktail (GE Healthcare, Germany) and purified by TALON beads (Clontech, USA).

Preparation of brain microvascular endothelial cells (BMEC)

Primary cultures of rat BMECs were prepared as described previously (Veszelka *et al.*, 2007), and plated on collagen type IV coated chamber slide (Sigma-Aldrich, USA). Cells were cultivated in DMEM (Sigma-Aldrich) supplemented with 20% plasma derived serum (First Link, UK), gentamicine, 2 mM L-glutamine, 100 µg/ml heparin (Sigma-Aldrich), 1 ng/ml basic fibroblast growth factor (Roche, Germany) and 4 µg/ml puromycin at 37 °C and 5% CO_2 protective atmosphere.

Activation of BMEC molecules by PilE4

BMEC monolayer was incubated with purified rPilE4 (0.25 µg/cm^2) for 24 hrs. BMECs not incubated with rPilE4 were served as a negative control. BMECs were harvested and total RNA was isolated with PureZOL RNA isolation reagent (Bio-Rad, USA). RNA was then treated with DNase I (Fermentas, USA), and reverse transcribed with RevertAid H minus First strand cDNA synthesis kit (Fermentas, USA). Quantitative PCR was performed to assess expression levels of selected genes (depicted in Figure 1). As a housekeeping gene, GAPDH was used. Gene expression was evaluated with $\Delta\Delta CT$ method with the help of iQ5 software (Bio-Rad, USA).

Results and discussion

From a total of 22 genes, 10 were significantly upregulated and 7 genes were downregulated. PilE4 did not alter gene expression for THBD, TLR-2, TLR-4, TRAF-6 and MMP-2 in BMECs. CD14 ($\Delta\Delta CT$ 0.42) and CD40 ($\Delta\Delta CT$ 0.53), CD molecules that activate inflammatory response, were upregulated. PECAM-1 ($\Delta\Delta CT$ 0.56), which involved in the integrins activation and leukocyte migration, was also upregulated. In contrast, PilE4 significantly decreased level of expression of adhesive molecules like ICAM-1 ($\Delta\Delta CT$ -1.19) followed by VCAM-1 ($\Delta\Delta CT$ -0.37) and CD80 ($\Delta\Delta CT$ -1.06). In contrast, expression of all interleukins tested in this study was downregulated (IL-1 $\Delta\Delta CT$ -1.13, IL-2 $\Delta\Delta CT$ -1.53 and IL-6 $\Delta\Delta CT$ -1.13), however TNF-α expression was evoked as in untreated BMECs ($\Delta\Delta CT$ 0.35). Among pattern recognition receptors only TLR6 ($\Delta\Delta CT$ 1.3) was upregulated. Induction of TLR6 was reflected in to the upregulation of MyD88 ($\Delta\Delta CT$ 0.61) and IRAK-1 ($\Delta\Delta CT$ 0.28). Surprisingly, NF-κβ was downregulated ($\Delta\Delta CT$ -0.53). The most significant gene upregulation among MMP family was of MMP-9 ($\Delta\Delta CT$ 3.7), while MMP-3 ($\Delta\Delta CT$ 1.29) and MMP-1 ($\Delta\Delta CT$ 0.61) were moderately upregulated. In contrast, level of VCAM-1 after PilE4 challenge was significantly downregulated, which correlates with the results published by Forestal and colleagues, where incubation of HUVEC with LVS led to downregulation of VCAM (Forestal *et al.*, 2003). Upregulated CD40 alters MMPs expression and vascular endothelial growth factor activation, what leads to the formation of fenestrations. Moreover, MMPs are responsible for degradation and remodelation of connective tissue, that causes brain damage (Pulzova *et al.*, 2011). In the PilE-treated BMECs, the expression of MMP-9 was the highest among studied

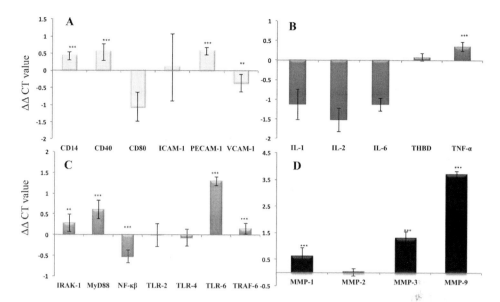

Figure 1. Panel A – expression of CD molecules and integrins in BMEC challenged with rPilE4; Panel B – expression of cytokines in BMEC challenged with rPilE4; Panel C – expression of IRAK-1, MyD88, NF-κβ, TRAF-6 and toll-like receptors in challenged BMEC with rPilE4; Panel D – expression of proteinases in BMEC challenged with rPilE4. Fold change in the expression was calculated using expression level normalized to the housekeeping gene GAPDH. Error bars indicate the standard deviation of pentaplicate samples. P values were calculated by using unpaired t test, comparing gene expression in BMECs infected with PilE4 of LVS and untreated BMECs (** P<0.02; *** P<0.0001).

matrix metalloproteases, followed by MMP-3 and MMP-1. Whereas expression of MMP-2 was rudimentory. Results indicate that *Francisella* may modulate ECM architecture through exploitation of MMP-9 for rapid dissemination from microvasculature in to the brain tissue. These events are necessary for the leukocyte adhesion and penetration through endothelium and also for the further activation of adhesion molecules and integrins. Results indicate that *Francisella* may modulate ECM architecture through exploitation of MMP-9 for rapid dissemination from microvasculature in to the brain tissue.

Acknowledgements

Financial support was from APVV-0036-10, VEGA – 2/0121/11. STSM mobility was funded through COST-Action FA1002 Farm animal proteomics. E.B., L.P. and Z.F. are funded by ITMS 26220220185.

References

Bencurova, E., Mlynarcik, P. and Bhide, M., 2011. An insight into the ligand-receptor interactions involved in the translocation of pathogens across blood-brain barrier. FEMS Immunol Med Microbiol 63: 297-318.

Bencurova, E., Mlynarcik, P., Pulzova, L., Kovac, A. and Bhide, M.R., 2013. PilE4 may contributes in the adhesion of Francisella to brain microvascular endothelial cells, Farm animal proteomics 2013. Springer, pp. 99-102.

Forestal, C.A., Benach, J.L., Carbonara, C., Italo, J.K., Lisinski, T.J. and Furie, M.B., 2003. Francisella tularensis selectively induces proinflammatory changes in endothelial cells. J Immunol 171: 2563-2570.

Lindemann, S.R., McLendon, M.K., Apicella, M.A. and Jones, B.D., 2007. An *in vitro* model system used to study adherence and invasion of Francisella tularensis live vaccine strain in nonphagocytic cells. Infect Immun 75: 3178-3182.

Melillo, A., Sledjeski, D.D., Lipski, S., Wooten, R.M., Basrur, V. and Lafontaine, E.R., 2006. Identification of a Francisella tularensis LVS outer membrane protein that confers adherence to A549 human lung cells. FEMS Microbiol Lett 263: 102-108.

Moreland, J.G., Hook, J.S., Bailey, G., Ulland, T. and Nauseef, W.M., 2009. Francisella tularensis directly interacts with the endothelium and recruits neutrophils with a blunted inflammatory phenotype. Am J Physiol Lung Cell Mol Physiol 296: L1076-1084.

Pulzova, L., Bhide, M.R. and Andrej, K., 2009. Pathogen translocation across the blood-brain barrier. FEMS Immunol Med Microbiol 57: 203-213.

Pulzova, L., Kovac, A., Mucha, R., Mlynarcik, P., Bencurova, E., Madar, M., Novak, M. and Bhide, M., 2011. OspA-CD40 dyad: ligand-receptor interaction in the translocation of neuroinvasive Borrelia across the blood-brain barrier. Sci Rep 1: 86.

Veszelka, S., Pasztoi, M., Farkas, A.E., Krizbai, I., Ngo, T.K., Niwa, M., Abraham, C.S. and Deli, M.A., 2007. Pentosan polysulfate protects brain endothelial cells against bacterial lipopolysaccharide-induced damages. Neurochem Int 50: 219-228.

Farm animal proteomics as a tool for understanding human health: the case of the ACOS-innovation Project

Luigino Calzetta[1]*, Mario Cazzola[1], Andrea Urbani[2], Paola Rogliani[1], Paola Roncada[3,4] and Luigi Bonizzi[3]

[1]*Department of Systems Medicine, University of Rome 'Tor Vergata', Rome, Italy;* luigicalz@gmail.com

[2]*Department of Experimental Medicine and Surgery, University of Rome 'Tor Vergata', Rome, Italy*

[3]*Dipartimento di Scienze Veterinarie e Sanità Pubblica (DIVET), Università degli Studi di Milano, Milan, Italy*

[4]*Istituto Sperimentale Italiano L. Spallanzani, Milan, Italy*

Asthma – Chronic Obstructive Pulmonary Disease (COPD) Overlap Syndrome (ACOS) affects up to 4.5% of general population and approximately 15-25% of patients with chronic obstructive airway diseases worldwide (De Marco *et al.*, 2013; Gibson and Simpson, 2009; Louie *et al.*, 2013; Nakawah *et al.*, 2013; Zeki *et al.*, 2011). ACOS is a mixed condition overlapping between asthma and COPD that is characterized by variable airflow obstruction, incomplete reversible airflow limitation and mixed inflammatory pattern. Although the relevant prevalence of this syndrome, to date neither randomized clinical trials have been carried out nor animal models have been proposed (Gibson and Simpson, 2009). ACOS patients are currently empirically treated with symptom-targeted approach by merging international guidelines for asthma and COPD, with no support from evidence-based medicine (Louie *et al.*, 2013). This scenario highlights a strong medical need for pioneering research in order to modelling and characterize ACOS and to perform studies on drug efficacy in a multidisciplinary context.

In this context, the ACOS-innovation Project provides the use of pig for modelling ACOS since this farm animal, from miniature to large breeds, is remarkably similar to humans in terms of anatomy, genetics, physiology and immune response. Specifically, mature miniature breeds have adult human-sized organs and, in contrast to mice and rats, the respiratory system has extensive interlobular and intralobular connective tissue as in humans. The immune system of swine is well characterized and offers a wide range of established methodologies and tools, most proteins of the immune system share structural and functional similarities (>80%) with their human counterparts, compared with rodents (<10%), and the immune response closely mimics that of human. Overall, considering that porcine immune system is functionally more similar to its human counterpart when compared to rodents, this model will be effective as translational approach to be predictive of therapeutic treatments of ACOS in humans, that results from host response overlapping between asthma and COPD. In addition, porcine models are cheaper and ethically more acceptable than primates (Meurens *et al.*, 2012). These evidences emphasize the adequacy of the proposed porcine model of ACOS since it will reproduce, after only three weeks of challenging, the main characteristics of both asthma (bronchial hyperresponsiveness) and COPD (emphysema, small airway remodelling/obstruction and bronchitis) in the same animal (Chen *et al.*, 2013; Turner *et al.*, 2001).

Therefore, the ACOS-innovation Project will propose to characterize and validate a novel, pioneering and exclusive animal model of ACOS, that will permit to characterize *in vivo* this disease in a translational setting for approaching adequate future human clinical trials. In addition, another aim of this Project will be to improve the knowledge of ACOS and, thus, to provide novel and evidence-based therapeutic approaches for this disease. In particular, the Project will permit to identify valuable pharmacotherapeutic and probiotic interventions and to discover suitable biomarkers in order to adequately investigate this syndrome and to assess the effectiveness of novel therapeutic interventions.

The ACOS model will be carried out in miniature pigs by inducing both asthma (ovalbumin sensitization) and COPD (lipopolysaccharide challenge) conditions in the same animal, in agreement with validated protocols recently proposed for each single model (Calzetta *et al.*, 2014; Chen *et al.*, 2013; Mitchell *et al.*, 1999; Smith *et al.*, 2011; Turner *et al.*, 2001, 2002).

The pulmonary function and the influence of probiotics administration will be tested *in vivo* and pharmacological experiments will be then performed on isolated airways of animals in which ACOS conditions have been induced. The best pharmacological interventions for improving the bronchial hyper-responsiveness and inflammation induced by ACOS will be investigated in both medium bronchi and bronchioles. This will include at the least a double and a triple drug combination, formulated with the first line current medications and/or the second line current medications and/or the main emerging therapeutic interventions for treating asthma and COPD. The drugs to be investigated will be the long-acting muscarinic antagonists (LAMA), the long-acting β_2 agonists (LABA), the inhaled corticosteroids (ICS), the 5-lypoxigenase (5-LO) inhibitors, the phosphodiesterase (PDE) 3 and 4 inhibitors and the mevalonate pathway inhibitors (statins) (Calzetta *et al.*, 2011, 2013; Cazzola *et al.*, 2011, 2012; Matera *et al.*, 2005, 2008, 2011).

After that, combined therapies will be also investigated in the *in vivo* model of ACOS, by analyzing also the effect of probiotics administration. In this phase of the Project, the influence of the proposed novel therapeutic approach for ACOS will be tested on the pulmonary function in order to validate results obtained from *ex vivo* settings (Calzetta *et al.*, 2014).

Tissues and samples will be collected during all the phases of the Project in order to perform biomarker discovery. In particular, biomarkers fingerprint profiling will be carried out by proteomic investigations, inflammatory findings and metaproteomic analysis of respiratory microbiome (Calabrese *et al.*, 2011; Del Chierico *et al.*, 2012, 2014; Piras *et al.*, 2014; Turk *et al.*, 2012; Urbani *et al.*, 2013). Figure 1 shows the Project algorithm by workflow.

Finally, the ACOS-innovation Project will provide novel farm animal approaches for biomarker discovery to investigate ACOS and improve therapy, via combining treatments in order to produce synergistic interaction between agents and, thus, reducing doses and adverse events. Therefore, this will be the first Project that will produce evidence-based findings concerning ACOS, a syndrome that has been always reductively considered a simple combination of asthma and COPD, and that

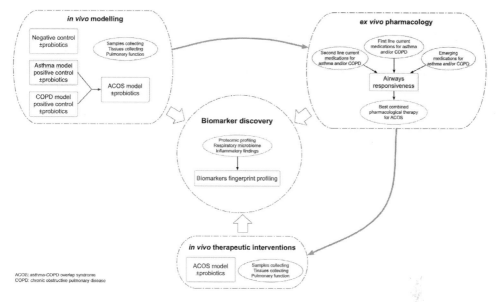

Figure 1. Project algorithm.

has been so far treated moving from empirical and indirect findings translated for asthma and COPD guidelines.

References

Calabrese, C., Marzano, V., Urbani, A., Lazzarini, G., Valerii, M.C., Liguori, G., Di Molfetta, S., Rizzello, F., Gionchetti, P., Campieri, M. and Spisni, E., 2011. Distinct proteomic profiles characterise non-erosive from erosive reflux disease. Alimentary Pharmacology and Therapeutics 34: 982-993.

Calzetta, L., Page, C.P., Spina, D., Cazzola, M., Rogliani, P., Facciolo, F. and Matera, M.G., 2013. Effect of the mixed phosphodiesterase 3/4 inhibitor RPL554 on human isolated bronchial smooth muscle tone. Journal of Pharmacology and Experimental Therapeutics 346: 414-423.

Calzetta, L., Rossi, P., Bove, P., Alfonsi, P., Bonizzi, L., Roncada, P., Bernardini, R., Ricciardi, E., Montuori, M., Pistocchini, E., Mauti, P. and Mattei, M., 2014. A Novel and Effective Balanced Intravenous-Inhalant Anaesthetic Protocol in Swine by Using Unrestricted Drugs. Experimental Animals.

Calzetta, L., Spina, D., Cazzola, M., Page, C.P., Facciolo, F., Rendina, E.A. and Matera, M.G., 2011. Pharmacological characterization of adenosine receptors on isolated human bronchi. American Journal of Respiratory Cell and Molecular Biology 45: 1222-1231.

Cazzola, M., Calzetta, L., Page, C.P., Rinaldi, B., Capuano, A. and Matera, M.G., 2011. Protein prenylation contributes to the effects of LPS on EFS-induced responses in human isolated bronchi. American Journal of Respiratory Cell and Molecular Biology 45: 704-710.

Cazzola, M., Calzetta, L., Rogliani, P., Lauro, D., Novelli, L., Page, C.P., Kanabar, V. and Matera, M.G., 2012. High glucose enhances responsiveness of human airways smooth muscle via the Rho/ROCK pathway. American Journal of Respiratory Cell and Molecular Biology 47: 509-516.

Chen, P., Hou, J., Ding, D., Hua, X., Yang, Z. and Cui, L., 2013. Lipopolysaccharide-induced inflammation of bronchi and emphysematous changes of pulmonary parenchyma in miniature pigs (Sus scrofa domestica). Lab Anim (NY) 42: 86-91.

De Marco, R., Pesce, G., Marcon, A., Accordini, S., Antonicelli, L., Bugiani, M., Casali, L., Ferrari, M., Nicolini, G., Panico, M.G., Pirina, P., Zanolin, M.E., Cerveri, I. and Verlato, G., 2013. The coexistence of asthma and chronic obstructive pulmonary disease (COPD): prevalence and risk factors in young, middle-aged and elderly people from the general population. PloS One 8: e62985.

Del Chierico, F., Petrucca, A., Vernocchi, P., Bracaglia, G., Fiscarelli, E., Bernaschi, P., Muraca, M., Urbani, A. and Putignani, L., 2014. Proteomics boosts translational and clinical microbiology. Journal of Proteomics 97: 69-87.

Del Chierico, F., Vernocchi, P., Bonizzi, L., Carsetti, R., Castellazzi, A.M., Dallapiccola, B., de Vos, W., Guerzoni, M.E., Manco, M., Marseglia, G.L., Muraca, M., Roncada, P., Salvatori, G., Signore, F., Urbani, A. and Putignani, L., 2012. Early-life gut microbiota under physiological and pathological conditions: the central role of combined meta-omics-based approaches. Journal of Proteomics 75: 4580-4587.

Gibson, P.G. and Simpson, J.L., 2009. The overlap syndrome of asthma and COPD: what are its features and how important is it? Thorax 64: 728-735.

Louie, S., Zeki, A.A., Schivo, M., Chan, A.L., Yoneda, K.Y., Avdalovic, M., Morrissey, B.M. and Albertson, T.E., 2013. The asthma-chronic obstructive pulmonary disease overlap syndrome: pharmacotherapeutic considerations. Expert Review of Clinical Pharmacology 6: 197-219.

Matera, M.G., Calzetta, L., Peli, A., Scagliarini, A., Matera, C. and Cazzola, M., 2005. Immune sensitization of equine bronchus: glutathione, IL-1beta expression and tissue responsiveness. Respiratory Research 6: 104.

Matera, M.G., Calzetta, L., Rogliani, P., Bardaro, F., Page, C.P. and Cazzola, M., 2011. Evaluation of the effects of the R- and S-enantiomers of salbutamol on equine isolated bronchi. Pulmonary Pharmacology and Therapeutics 24: 221-226.

Matera, M.G., Calzetta, L., Sanduzzi, A., Page, C.P. and Cazzola, M., 2008. Effects of neuraminidase on equine isolated bronchi. Pulmonary Pharmacology and Therapeutics 21: 624-629.

Meurens, F., Summerfield, A., Nauwynck, H., Saif, L. and Gerdts, V., 2012. The pig: a model for human infectious diseases. Trends in Microbiology 20: 50-57.

Mitchell, H.W., Turner, D.J., Gray, P.R. and McFawn, P.K., 1999. Compliance and stability of the bronchial wall in a model of allergen-induced lung inflammation. J Appl Physiol (1985) 86: 932-937.

Nakawah, M.O., Hawkins, C. and Barbandi, F., 2013. Asthma, chronic obstructive pulmonary disease (COPD), and the overlap syndrome. Journal of the American Board of Family Medicine 26: 470-477.

Piras, C., Soggiu, A., Greco, V., Cassinotti, A., Maconi, G., Ardizzone, S., Amoresano, A., Porro, G.B., Bonizzi, L. and Roncada, P., 2014. Serum protein profiling of early and advanced stage Crohn's disease. EuPA Open Proteomics.

Smith, K.R., Leonard, D., McDonald, J.D. and Tesfaigzi, Y., 2011. Inflammation, mucous cell metaplasia, and Bcl-2 expression in response to inhaled lipopolysaccharide aerosol and effect of rolipram. Toxicology and Applied Pharmacology 253: 253-260.

Turk, R., Piras, C., Kovacic, M., Samardzija, M., Ahmed, H., De Canio, M., Urbani, A., Mestric, Z.F., Soggiu, A., Bonizzi, L. and Roncada, P., 2012. Proteomics of inflammatory and oxidative stress response in cows with subclinical and clinical mastitis. Journal of Proteomics 75: 4412-4428.

Turner, D.J., Gray, P.R., Taylor, S.A., Thomas, J. and Mitchell, H.W., 2001. Physiological responses of the airway wall and lung in hyperresponsive pigs. European Respiratory Journal 18: 935-941.

Turner, D.J., Noble, P.B., Lucas, M.P. and Mitchell, H.W., 2002. Decreased airway narrowing and smooth muscle contraction in hyperresponsive pigs. J Appl Physiol (1985) 93: 1296-1300.

Urbani, A., De Canio, M., Palmieri, F., Sechi, S., Bini, L., Castagnola, M., Fasano, M., Modesti, A., Roncada, P., Timperio, A.M., Bonizzi, L., Brunori, M., Cutruzzola, F., De Pinto, V., Di Ilio, C., Federici, G., Folli, F., Foti, S., Gelfi, C., Lauro, D., Lucacchini, A., Magni, F., Messana, I., Pandolfi, P.P., Papa, S., Pucci, P., Sacchetta, P. and Italian Mt-Hpp Study Group-Italian Proteomics, A., 2013. The mitochondrial Italian Human Proteome Project initiative (mt-HPP). Mol Biosyst 9: 1984-1992.

Zeki, A.A., Schivo, M., Chan, A., Albertson, T.E. and Louie, S., 2011. The Asthma-COPD Overlap Syndrome: A Common Clinical Problem in the Elderly. J Allergy (Cairo) 2011: 861926.

Proteomics of the mitochondrial proteome in dairy goats (*Capra hircus*)

Graziano Cugno[1,2], Lorenzo E. Hernandez-Castellano[2,3], Mariana Carneiro[1], Noemí Castro[2], Anastasio Argüello[2], Juan Capote[4],Sébastien Planchon[5], Jenny Renaut[5], Alexandre M. Campos[1] and André M. Almeida[6*]*

[1]*CIMAR/CIIMAR, Centro Interdisciplinar de Investigação Marinha e Ambiental, Universidade do Porto, Porto, Portugal*

[2]*Department of Animal Science, Universidad de Las Palmas de Gran Canaria, Arucas, Spain*

[3]*Veterinary Physiology, Vetsuisse Faculty, University of Bern, Bern, Switzerland*

[4]*ICIA, Instituto Canario de Investigaciones Agrarias, Valle Guerra, Tenerife, Spain*

[5]*Centre de Recherche Publique Gabriel Lippmann, Belvaux, Luxemburg*

[6]*IBET, Instituto de Biologia Experimental e Tecnológica, Oeiras, Portugal, IICT – Instituto de Investigação Científica Tropical, Lisboa, Portugal, CIISA-Centro Interdisciplinar de Investigação em Sanidade Animal, Lisboa, Portugal and Instituto de Tecnologia Química e Biológica António Xavier da Univ. Nova de Lisboa, Oeiras, Portugal; aalmeida@fmv.utl.pt*

Introduction

Goat milk production is important in the EU, being Italy, France, Spain and Portugal the main producers of the commercialized goat milk and cheese. Nevertheless seasonal weight loss (SWL) poses limitations to animal production in Tropical and Mediterranean regions, conditioning producer's incomes and the nutritional status of rural communities. It is of the utmost importance to produce strategies to oppose adverse effects of SWL. Breeds that have evolved in harsh climates have acquired a tolerance to SWL through selection. Most of the factors determining such ability are related to biochemical metabolic pathways and are likely important biomarkers to SWL. In this study, a gel based proteomics strategy (BN: Blue-Native Page and 2DE: two-dimensional gel electrophoresis) was used to characterize the mitochondrial proteome of the secretory tissue of the caprine mammary gland. In addition, we have also conducted an investigation of the effects of weight loss in two dairy goat breeds with different levels of adaptation to nutritional stress: Majorera (tolerant) and Palmera (susceptible) from the Canary Islands (Spain).

Material and methods

Experimental design and sample collection: the study was conducted using 10 Majorera and 10 Palmera dairy goats, divided in 4 sets, 2 for each breed: underfed group fed on wheat straw ad libitum (restricted diet, so their body weight would be 15-20% reduced by the end of experiment), and a control group fed ad libitum on commercial feed (Lérias *et al.*, 2013). After 22 days, mammary gland biopsies on the animals were conducted following standard procedures and under competent veterinary supervision.

Isolation of Mitochondria: tissues were disrupted in liquid nitrogen and homogenized in sucrose (0.35 M), EDTA (1.5 M), Tris (1 mM) and BSA (1%, w/v), pH 7.4. Homogenates were first centrifuged for 10 min, at 700×g and 4 °C and afterwards at 8,800×g, 10 min and the mitochondria pellet retained.

Blue-Native PAGE: mitochondrial protein complexes were solubilized at 1.4 mg/ml, in aminocaproic acid (ACA) (750 mM), BisTris (50 mM), Na-EDTA (0.5 mM), n-Dodecyl-beta-D-maltoside (DDM) (1:7 detergent:protein ratio, w/w), pH 7.0. A negative charge was conferred to proteins with Coomassie Blue G-250 and proteins were subsequently separated in native acrylamide gels (5%-13% acrylamide gradients) (Campos *et al.*, 2010). BN-PAGE lanes were cut and protein complexes denatured in gel to enable a second separation in gel (SDS-PAGE) (Campos *et al.*, 2010).

2 Dimension Electrophoresis: mitochondrial proteins were solubilized with urea (7 M), thiourea (2 M), CHAPS (4%, w/v), DTT (65 mM), ampholytes (0.8%, v/v) and separated by 2DE (Campos *et al.*, 2010). Gels were subsequently stained with fluorescent dye (Oriole, Biorad, Hercules, USA) and analysed with PDQuest image software (Biorad).

Protein identification: proteins were isolated and in-gel digested with trypsin. Peptides were analysed in a 5800 Proteomics Analyzer (ABsciex, Framingham, USA) (Printz *et al.* (2013). Protein identification was done by searching the MS and MS/MS data against NCBI database in the Other Mammalia taxonomy (434586 sequences), two trypsin missed cleavages, carbamidomethylation of cysteine as fixed modification as well as four dynamic modifications (methionine and tryptophan oxidation, tryptophan dioxydation and tryptophan to kynurenin) were allowed. Mass accuracy was set to 100 ppm for parent ions and 0.5 Da for MS/MS fragments. Homology identification was retained with probability set at 95%. Protein classification with gene ontology terms was performed in the program Panther (http://www.pantherdb.org) with the human protein homologs.

Results and discussion

The proteomic analysis of the mitochondria of mammary glands, upon organelle isolation, enabled the resolution of a total of 277 proteins, being 184 (66%) of them identified by MALDI-TOF/TOF mass spectrometry. Among the identified proteins, we could detect several subunits of the glutamate dehydrogenase complex and the respiratory complexes I, II, IV, V from mitochondria (Figure 1). From the 2DE analysis we were able to identify and classify proteins according to the biological processes they are involved in, particularly cell metabolism, development, localization, cellular organization and biogenesis, biological regulation, response to stimulus, to cite only the most relevant (Figure 1). These proteins were mapped in both BN and 2DE gels. The comparative proteomics analysis enabled the identification of Succinyl-CoA synthetase, Guanine nucleotide-binding protein, NADH-ubiquinone oxidoreductase, in *Majorera*, and ACTA2 protein in *Palmera*, as being over-expressed as a consequence of weight loss.

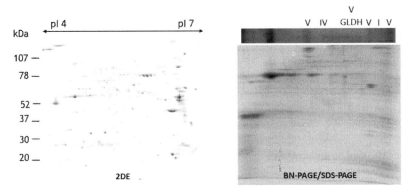

Figure 1. Characterization of the goat mammary gland mitochondrial proteome, using two-dimensional gel electrophoresis and Blue-Native PAGE.

Conclusion

Our results showed that it is possible to use this protocol to analyze mitochondrial proteomes of ruminant mammary gland species, setting the goat mammary gland as reference. Future studies will focus the increase of the coverage of the mitochondrial proteome and in the identification of additional markers with roles in the adaptation of goats to weight loss, lactation and milk production.

Acknowledgements

This project was funded by the research project PTDC/CVT/116499/2010 (FCT, Lisbon, Portugal) and by the Cost Action FA 1002 through a Short-term Scientific Mission. Authors acknowledge the participation of Dr A. Morales de la Nuez, P. Delgado del Olmo, D. Martell-Jaizme and A. Suárez-Trujillo during the sample collection.

References

Campos A, Carvajal-Vallejos PK, Villalobos E, Franco CF, Almeida AM, Coelho AV, Torné JM and Santos M, 2010. Characterization of Zea mays L. plastidial transglutaminase: interactions with thylakoid membrane proteins. Plant Biol 12, 708-16.

Lérias JR, Hernández-Castellano LE, Morales-Delanuez A, Araújo SS, Castro N, Argüello A, Capote J and Almeida AM, 2013. Body live weight and milk production parameters in the Majorera and Palmera goat breeds from the Canary Islands: influence of weight loss. Trop Anim Health Prod. 45, 1731-6.

Printz B, Sergeant K, Guignard C, Renaut J and Hausman JF, 2013. Physiological and proteome study of sunflowers exposed to a polymetallic constraint. Proteomics 13, 1993-2015.

A proteomic study of bovine whey in a model of Gram positive and Gram negative bacterial mastitis

Funmilola C. Thomas[1]*, Timothy Geraghty[2], Patricia B.A. Simoes[2], Lorraine King[3], Richard Burchmore[3] and Peter D. Eckersall[1]

[1]Institute of Biodiversity, Animal Health & Comparative Medicine, University of Glasgow, 464 Bearsden Road, Glasgow, G61 1QH, United Kingdom; f.thomas.1@research.gla.ac.uk

[2]School of Veterinary Medicine, University of Glasgow, 464 Bearsden Road, Glasgow, G61 1QH, United Kingdom

[3]Glasgow Polyomics; University of Glasgow, 464 Bearsden Road, Glasgow, G61 1QH, United Kingdom

Introduction

Mastitis continues to be a major economic and health burden to the dairy industry worldwide (Halasa *et al.*, 2007). Major pathogens which have been incriminated in the aetiology of bovine mastitis include Gram negative and Gram positive bacteria. Environmental pathogens such as *Escherichia coli*; a Gram negative coliform bacterium and *Streptococcus uberis* a gram positive bacterium have been associated with severe forms of bovine mastitis and show high prevalence rates in bovine mastitis worldwide (Bradley *et al.*, 2007; Unnerstad *et al.*, 2009). In recent years, milk proteomics has gained importance in identifying protein changes in relation to bovine mastitis (Ceciliani *et al.*, 2014). A number of researchers have utilized the proteomics technology in order to identify potential biomarkers of bovine mastitis in milk samples (Roncada *et al.*, 2012). The outcome of these studies is that new proteins not previously reported in milk have been identified in both normal and mastitic whey. To further explore the potentials of milk proteomics, a liquid phase isoeletric fractionation step has been employed for pre-fractionation of proteins in whey of mastitis versus healthy bovine milk. Pooled whey samples from *E. coli* and *S. uberis* (selected as a model of gram negative and gram positive bacteria mastitis respectively) along with healthy milk were examined in this study.

Materials and methods

High speed centrifugation ($3,500 \times g$ for 30-60 minutes at 4 °C) was used in sample preparation step to obtain milk whey from pools of *E. coli*, *S. uberis* as well as healthy milk udders. The milk samples were from natural cases of mastitis submitted for bacteriological examination to the Veterinary Diagnostic Services (School of Veterinary Medicine, University of Glasgow, UK) from 7 dairy farms across Scotland between August 2012 and December 2013. From the results of the bacterial culture and isolation, aliquots of all samples positive for *E. coli* (n=9) and *S. uberis* n=15) (pure cultures) were pooled. Milk samples from bacteriologically negative milk samples with somatic cell counts (SCC) ≤100,000 cells/ml (n=12) were also aliquoted into a pooled sample. Pool samples were dialyzed, filtered and had their protein concentration determined. A mini Rotofor® system (220/240 V, 18 ml sample volume; Bio-Rad Laboratories, Hemel Hempstead, UK) was used to fractionate each whey sample into *pI* fractions. The preparative isoelectric focusing was carried out according to the manufacturer's protocol and the whey was separated into 20 fractions under constant power of 12 W for 3 hours. The pH of each fraction was determined, and fractions were then dialyzed and concentrated, and protein concentration determined. Fractions were then resolved by molecular weight using 1DE SDS-PAGE on 5-14% precast Ready-gel (Bio-Rad, Bio-Rad Labs, Hemel Hempstead, UK) and then stained with Coomasie blue, bands of interest were excised digested with trypsin and a liquid chromatography tandem mass spectrometry (LC-MS/MS) was carried out for protein identification. Proteomics experiment was carried out at the Glasgow-Polyomics facility of the University of Glasgow.

Results and discussion

Scanned images of 1DE gels showing excised bands (circled) and their identification in each pool fractions are shown in Figure 1, 2 and 3. There were some differences between the pools of milk samples and compared to normal 2DE gels, some of these differences are highlighted in squares on the figures. Compared with using a gel based isoelectric focusing of proteins, it can be concluded that liquid phase isoelectric focusing could offer an advantage of concentrating low abundance proteins away from the high abundance proteins (caseins and major whey proteins) at their *pI* of 4-5, thus making these lower abundance proteins easier to identify.

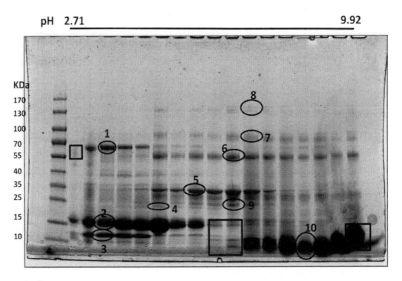

Figure 1. Rotofor® fractions from pool of healthy bovine whey resolved on 1DE.

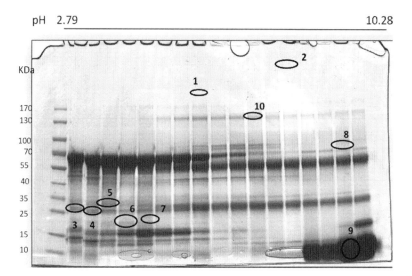

Figure 2. Rotofor® fractions from pool of E. coli *infected bovine whey resolved on 1DE.*

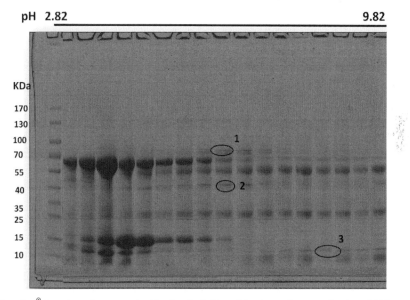

Figure 3. Rotofor® fractions from pool of S. uberis *infected bovine whey resolved on 1DE.*

Acknowledgements

Funding from the Tertiary Education Trust Fund Nigeria and Federal University of Agriculture Abeokuta for FCT is gratefully acknowledged.

References

Bradley, A., Leach, K., Breen, J., Green, L. and Green, M., 2007. Survey of the incidence and aetiology of mastitis on dairy farms in England and Wales. Veterinary Record 160: 253-258.

Ceciliani, F., Restelli, L. and Lecchi, C., 2014. Proteomics in farm animals models of human diseases. Proteomics Clin. Appl. 8: 677-688.

Halasa, T., Huijps, K., Osteras, O. and Hogeveen, H., 2007. Economic effects of bovine mastitis and mastitis management: A review. Veterinary Quarterly 29: 18-31.

Roncada, P., Piras, C., Soggiu, A., Turk, R., Urbani, A. and Bonizzi, L., 2012. Farm animal milk proteomics. Journal of Proteomics 75: 4259-4274.

Unnerstad, H., Lindberg, A., Waller, K., Ekman, T., Artursson, K., Nilsson-Ost, M. and Bengtsson, B., 2009. Microbial aetiology of acute clinical mastitis and agent-specific risk factors. Veterinary Microbiology 137: 90-97.

Blood haptoglobin concentration in New Zealand white rabbits during first year of their life

Teodora M. Georgieva[1]*, Vladimir S. Petrov[2], Aanna Bassols[3], Evgenya V. Dishlyanova[1], Ivan P. Georgiev[1], Kalina N. Nedeva[1], Mariela I. Koleva[1], Fabrizio Ceciliani[4] and Tatyana Vlaykova[5]

[1]Department of Pharmacology, Animal Physiology and Physiological Chemistry, Faculty of Veterinary Medicine, Trakia University, 6 000 Stara Zagora, Bulgaria; teodoramirchevag@abv.bg
[2]Department of Veterinary Medical Microbiology, Infection and Parasitic Diseases, Faculty of Veterinary Medicine, Trakia University, 6 000 Stara Zagora, Bulgaria
[3]Department de Bioquimica i Biologia Molecular, Facultat de Veterinaria, Universitat Autronoma de Barcelona, Spain
[4]Dipartimento di Scienze Animali e Salute Pubblica, Università degli Studi di Milano, Milano, Italy
[5]Department of Chemistry and Biochemistry, Faculty of Medicine, Trakia University, 6 000 Stara Zagora, Bulgaria

Objectives

The acute phase response (APR) refers to a nonspecific and complex reaction of an animal that occurs shortly after any tissue injury. The origin of the response can be attributable to infectious, immunologic, neoplastic, traumatic, or other causes, and the purpose of the response is to restore homeostasis and to remove the cause of disturbance (Ceron *et al.*, 2005; Whicher and Westacott, 1992).

The production and secretion by the liver of a number of acute phase proteins (APP), mainly glycoproteins, is one of the mechanisms involved in the response to mediators produced by leucocytes and macrophages during episodes of infection or inflammation (Eckersall and Conner, 1988). In animals, several classifications have been proposed on the basis of the human model for post infection APP behaviour. Haptoglobin (Hp), ceruloplasmin and fibrinogen were affirmed to be the major positive APPs in rabbits (Petersen *et al.*, 2004).

According to cited researchers, Hp increased more than 10 times in rabbits, while fibrinogen – between 2 and 10 times. Contrary to this statement, Gruys *et al.* 1993 reported that in rabbits, the role of Hp was not so important, but instead, CRP and SAA were the major APPs. Recently, Theilgaard and Monch have identified Hp as a protein within the granules of neutrophils (Theilgaard-Monch *et al.*, 2006).

To the best of our knowledge there is no available information about the concentration of Hp in normal male and female New Zealand rabbits at different ages, except the data of Georgieva *et al.* at age of 3, 3.5 and 4 months in obese male rabbits (Georgieva *et al.*, 2011). According to some directives of the Society of Clinical Pathology for scientific purposes it could be good to know the normal ranges of Hp in rabbits and to used them for comparison with the ranges of Hp in all diseases with inflammation, including infection.

Therefore the aim of this study was to investigate the effect of age, gender and technological regimen on plasma concentration of haptoglobin in rabbits during first year of their life.

Material and methods

Animals and experimental design

The experimental procedure was approved by the Ethic Committee at the Faculty of Veterinary Medicine. Experiments were carried out with 12 New Zealand white rabbits divided in 2 groups: I^{st} group- 6 male and IInd group 6 female at ages of one month. Animas were followed until the 7th month, when female were pregnant and males were replacement rabbits having attained maturity. All animals were born from healthy doe rabbits and kept in the rabbitry. They were fed with pelleted feed according to their age and have had free access to the tap water.

Blood samples from assessing of haptoglobin were drown with heparin from *v. auricularis externa* at age 1, 2, 3, 6 and 7 months and plasma was obtained and stored at -20 °C until the assay.

Biochemical analysis

Hp concentration was measured in plasma using the kit of Tridelta, in which haptoglobin from samples was incubated with haemoglobin (Hb) to form an Hp-Hb complex. The pH was then reduced by addition of chromogene which destroys the peroxidase activity of the unbound Hb. The mixture generates hydrogen peroxide which causes a blue colour which was measured spectrophotometrically on Biochemical analyser Olympus at the Faculty of Veterinary Medicine in Barcelona. The preservation of the peroxidase activity of the Hb is directly proportional to the amount of Hp present in the sample. By comparison to standards with known concentrations of Hp the assay is calibrated and samples with unknown Hp concentration measured. The concentrations of Hp of the animals were compared between the genders and between the different periods of the experiment.

Statistical analysis

The statistical analysis of the data was performed using SPSS 16.0 for Windows (SPSS Inc.). Difference of Hp between genders was evaluated by Student t-test, while the differences of Hp in the groups between the beginning (month 1) and later periods were assessed by paired *t*-test. All data were expressed as mean ± standard deviation (SD) and the differences were considered significant when $P<0.05$.

Results and discussion

As shown in Figure 1, in male rabbits the mean plasma Hp concentration increased almost twice at month 2 (0.35±0.11 mg/ml), and at age of months 3 and 6 it increased approximately 6-fold (1.16±0.60 and 1.20±0.56 mg/ml, respectively, $P<0.001$) compared with the level of Hp at the age of month 1 (0.19±0.03 mg/ml). At the age of month 7 the values decreased (0.81± 0.26 mg/ml) but remained significantly higher than those of the beginning (month 1, $P<0.01$).

At the initial measuring time (1 month), the mean Hp concentration in female rabbits was 0.18±0.06 mg/ml and was comparable to that in the male rabbits. Then the kinetics of the Hp concentration followed the same pattern as it was in male rabbits, however the increased was with smaller magnification: the increases at age of months 3 and 6 were 3-fold compared with that at the age of month 1 ($P<0.05$) (Figure 1). There were significantly higher values in male compared to female rabbits only at the age of months 3 and 6 ($P<0.01$): the mean value of Hp at age of month 6 in female rabbits was twice lower than that in male rabbits. The mean value of Hp in pregnant female rabbits, what happened at age of 7 months, continued to increase and reached the value of 0.64±0.17 mg/ml.

The major finding of this study is that Hp concentration varies with the age and is affected by gender. This may contribute to further elucidation of Hp behavior in response to infection and various tissue damages. The relatively limited number of studies has shown that regardless of some variations, the development of acute phase response in rabbits is comparable to that in humans (Gruys, 2004; Kostro *et al.*, 2009; Mackiewicz *et al.*, 1988).

Serum APPs concentrations are related to the extent of disorders and the presence of tissue damage in affected animals, therefore, the alterations in their levels could be of diagnostic and prognostic

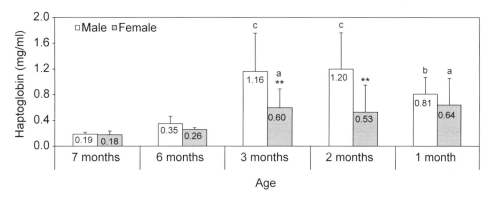

*Figure 1. The levels of haptoglobin at different ages in male and female animals. Data are presented as mean±SD (^{a}P<0.05, ^{b}P<0.01, ^{c}P<0.001 – significance of differences between the initial measurement at month 1 and later periods in both groups; ** P<0.01 – significance of differences between male and female animals in particular periods).*

value, if samples are collected at the appropriate time (Murata *et al.*, 2004). A similar response is established in animals, although in the various species it could vary substantially. Thus, Barsanti *et al.*, (1977) and Liberg (1977) established increase blood α- and β- globulin concentrations in dogs and cattle respectively.

While CRP and SAA are considered as major APPs in rabbits the available literature does not present convincing data for haptoglobin in rabbits (Gruys, 2004; Mackiewicz *et al.*, 1988; Petersen *et al.*, 2004).

In this respect, the results of the current investigation provide the lacking information about haptoglobin concentrations according to age, sex and the production cycle stage: suckling (1 month of age), 2 and 3 months of age, which are further divided in fattening and growing rabbits; 6 months of age – replacement rabbits having attained maturity and breeding rabbits – 7 months of age and older. The changes of haptoglobin during the acute phase response to bacterial infections is poorly elucidated, while there is no study at all in viral infection APR (Petersen *et al.*, 2004).

To the best our knowledge, this is the first work aiming to investigate Hp concentrations at different age and to compare the levels between the two genders and among rabbits in the different production groups. Hence, the current results could serve as a background for comparison of normal haptoglobin concentrations in healthy rabbits and to interpret the altered levels in response to various infections.

References:

Barsanti, J.A., Kristensen, F. and Drumheller, F.B., 1977. Analysis of serum proteins, using agarose electrophoresis in normal dogs and in dogs naturally infected with Dirofilaria immitis. Am J Vet Res 38: 1055-1058.

Ceron, J.J., Eckersall, P.D. and Martynez-Subiela, S., 2005. Acute phase proteins in dogs and cats: current knowledge and future perspectives. Vet Clin Pathol 34: 85-99.

Eckersall, P.D. and Conner, J.G., 1988. Bovine and canine acute phase proteins. Vet Res Commun 12: 169-178.

Georgieva, T., Bassols, A., Dishlyanova, E., Petrov, V., Marutsov, M., Dinev, I. and Georgiev, I., 2011. Blood haptoglobine response in obese rabbits with experimentally induced St. aureus, COST-Farm Animal Proteomics Spring Meeting 2011, Glasgow, pp. 49.

Gruys, E., 2004. Protein folding pathology in domestic animals. J Zhejiang Univ Sci 5: 1226-1238.

Gruys, E., van Ederen, A.M., Alsemgeest, S.P.M., Kalsbeek, H.C. and Wensing, T., 1993. Acute phase protein values in blood of cattle as indicator of animals with pathological processes. Archiv Lebensmittelhyg 44: 107-113

Kostro, K., Jarosz, Gruszecki, T., Junkuszew, A. and Lipecka, C., 2009. Utility of haptoglobin assay for sheep welfare and health status evaluation in pre- and post-slaughter period. Bull. Vet. Inst. Pulawy 53: 111-116.

Liberg, P., 1977. Agarose gel electrophoretic fractionation of serum proteins in adult cattle. II. A study of cows with different diseases. Acta Vet Scand 18: 335-348.

Mackiewicz, A., Ganapathi, M.K., Schultz, D., Samols, D., Reese, J. and Kushner, I., 1988. Regulation of rabbit acute phase protein biosynthesis by monokines. Biochem J 253: 851-857.

Murata, H., Shimada, N. and Yoshioka, M., 2004. Current research on acute phase proteins in veterinary diagnosis: an overview. Vet J 168: 28-40.

Petersen, H.H., Nielsen, J.P. and Heegaard, P.M., 2004. Application of acute phase protein measurements in veterinary clinical chemistry. Vet Res 35: 163-187.

Theilgaard-Monch, K., Jacobsen, L.C., Nielsen, M.J., Rasmussen, T., Udby, L., Gharib, M., Arkwright, P.D., Gombart, A.F., Calafat, J., Moestrup, S.K., Porse, B.T. and Borregaard, N., 2006. Haptoglobin is synthesized during granulocyte differentiation, stored in specific granules, and released by neutrophils in response to activation. Blood 108: 353-361.

Whicher, J. and Westacott, C., 1992. The acute phase response. In: J. Wicher and S. Evans (Eds.), Biochemistry of Inflammation. Kluwer Academic, London, pp. 243-272.

A comparative proteomic analysis of human umbilical vein endothelial cells after infection with the rodent-borne hemorrhagic fevers puumala hantavirus and *Leptospira interrogans* serovar Copenhageni

Marco Goeijenbier[1][*], Byron E.E. Martina[1,2], Marga G.A. Goris[3], Ahmed Ahmed[3], Rudy Hartskeerl[3], Sebastién Planchon[4], Kjell Sergeant[4], Jenny Renaut[4], Eric C.M. van Gorp[1], Jarlath Nally[5,6] and Simone Schuller[5,7]

[1]*Department of Viroscience laboratory, Erasmus MC, room ee1671, Rotterdam, CE 50 3015, the Netherlands; m.goeijenbier@erasmusmc.nl*
[2]*ARTEMIS One Health Research Institute, Utrecht, the Netherlands*
[3]*Royal Tropical Institute (KIT), KIT Biomedical Research, Amsterdam, the Netherlands*
[4]*Gabriel Lipmann institute, Luxembourg city, Luxembourg*
[5]*Veterinary faculty, University College Dublin, Dublin, Ireland*
[6]*Bacterial Diseases of Livestock Research Unit, National Animal Disease Center, Agricultural Research Service, United States Department of Agriculture, Ames, IA 50010, USA*
[7]*Veterinary faculty, University of Bern, Bern, Switzerland*

Background

Pathogenic *Leptospira* and hantaviruses are pathogens with highly similar features regarding epidemiology, clinical manifestations and the fact that they are both highly neglected diseases (Goeijenbier *et al.*, 2014). Both pathogens are able to cause a disease which is mainly characterized by renal failure, (severe) hemorrhage and even death (Bharti *et al.*, 2003; Goeijenbier *et al.*, 2013). Despite the high burden of disease in humans Leptospirosis is also a major problem in veterinary medicine with many infections and high mortality rates in various mammalian species (Adler and de la Pena, 2010). Furthermore Leptospirosis causes an increased abortion rate in cattle and livestock. The die-off and subsequent decrease in reproduction have a major impact on the agricultural economy and could have massive impacts on individual farmers (Ellis, 1994; Hamond *et al.*, 2013). For both pathogens the way they cause disease, and especially the way they cause hemorrhage and renal failure, remains largely unknown. Based on *in vitro* and animal *in vivo* data there is a strong suggestion endothelial cells play a pivotal role in the pathogenesis of both diseases. For leptospirosis this assumption is supported by clinical data in severe leptospirosis patients from Indonesia who showed extreme activation of coagulation, in many cases followed by (fatal) hemorrhage (Chierakul *et al.*, 2008; Wagenaar *et al.*, 2010). *In vitro* work showed the ability of Leptospira to transmigrate through endothelial cell monolayers. Furthermore, outer membrane proteins and killed Leptospira cause up-regulation of several endothelial activation like I-CAM and E-Selectin suggesting alterations in the functionality of the vascular endothelium (Gomez *et al.*, 2008; Martinez-Lopez *et al.*, 2010; Vieira *et al.*, 2007).

Hantaviruses directly infect and replicate in endothelial cells which is proven both *in vitro* and *in vivo*. Without causing any direct damage to the endothelial cells there is a strong suggestion that the infection does hamper the function of endothelial cells (Mackow and Gavrilovskaya, 2009). For instance hantavirus infected endothelial cells have a decreased migratory function *in vitro*, adhere quiescent platelets and, in combination with VEGF, show an increase in permeability (Gavrilovskaya *et al.*, 2002; Gavrilovskaya *et al.*, 2010; Taylor *et al.*, 2013).

To further study the factors involved in the changes in the vascular endothelium in both pathogens we chose a comprehensive, discovery based, proteomic approach.

Objective

Generate a 'bird-eyes view' on the proteome of infected endothelial cells with PUUV and *Leptospira interrogans* serovar Copenhageni. By combining the results from both the bacterial infection and the hantavirus infection we aimed to especially identify proteins specific for pathognomic mechanisms instead of the many acute phase proteins seen in either bacterial or viral disease.

Methods

Leptospira and hantaviruses are known to heavily adapt under *in vitro* culture conditions. Therefore we have put substantial effort in isolating the pathogens used in this experiment while remaining their virulence. Puumala (PUUV) hantavirus was isolated from chronically infected bank voles (*Myodes glareolus* and the *Leptospira interrogans* serovar Copenhageni (RJ16441) was isolated from blood cultures of a clinically infected human. The virulent state of the isolate was confirmed in a guinea pig LPHS model, culture passages were kept to a minimum.

In a 24 wells plate 2.6×10^5 primary cultured endothelial cells were infected with a multiplicity of infection (MOI) of 3 either with PUUV or *Leptospira*. Cells and supernatant were used to confirm and monitor the infection of the endothelial cells. At 48 hours post infection, when 90-100% of the cells were infected, monolayer of cells were washed three times with 10 mM Tris 10 mM EDTA buffer (PH 8). Cells were solubilised by incubation with 100 µl solubilisation buffer (7 M urea, 2 M thiourea, 1% ASB-14) overnight until all cells were lysed and protein were solubilised. The solubilised cells were used for 1-Dimensional (1-D) Gel Electrophoresis, 2-Dimensional (2-D) Gel Electrophoresis and eventual 2-D DIGE. The differentially expressed protein spots in the 2-D- DIGE experiment were excised from the 2-D DIGE large cast gels and digested with trypsin using the fully automated Ettan Spot handling Workstation (GE Healthcare). After digestion, the peptides were solubilised and 0.7 µl of the peptide mixture spotted on a MALDI target plate. A total of 1 µl of matrix solution (alpha cyano-4-hydroxycinnamic acid in 50% ACN/0.1% TFA) was aded. All MS and MS/MS analyses were performed using a 5800 MALDI TOF/TOF (Applied Biosystems, Foster City, CA, USA). Protein identifications were imported into IPA ® and mapped to items in the Ingenuity ® knowledge database. Furthermore Pubmed/MEDLINE was searched to further identify potential interactions between the identified proteins and hantaviruses, Leptospira or endothelial cell functioning.

Results and discussion

To our knowledge this study is the first to report proteome changes of human endothelial cells after infection with PUUV hantavirus or pathogenic *Leptospira*. Confirmation and analysis of the infection kinetics revealed some interesting observations. Our low passage PUUV isolates efficiently infected and replicated in the HUVEC primary cell culture. Immune peroxidase staining showed the presence of intracellular PUUV nucleoprotein with the amount of infected cells reaching 100% after 48 hours. Interestingly, the DNA of the *Leptospira*, as quantified by qRT-PCR, seemed to also increase in the cell lysate of the HUVECs. Taken in mind that there was a clear intracellular staining positive for the presence of *Leptospira* serovar Copenhageni suggests the ability of *Leptospira* to move intracellular, something that has been suggested before, and possibly even intracellular replication. Further *in vitro* and *in vivo* experiments with non-virulent and maybe even non-pathogenic *Leptospira* are necessary to fully understand our observations, for this study it was important to prove that the *Leptospira* were able to interact with the HUVEC monolayer during the incubation period. Evaluation of the proteome of the lysed HUVEC's separated by molecular weight by 1D electrophoresis, showed us that no large obvious differences were present after 48 hours of infection. There were no visibly detectable differences between the lanes especially no extra bands appearing or any bands disappearing between the three conditions. However, two dimensional analysis revealed many proteins with a changed abundance in the different infection conditions. After protein separation, with subsequent DIGE labelling, differences in protein abundance was determined this resulted in 160 spots for PUUV vs Control, 150 for *Leptospira* versus control and 79 spots if data from both infections was combined. Interestingly no viral or Leptospiral proteins were identified in the fraction of infected cells, however a substantial part of the identified proteins were from bovine origin, which were present in the cell culture media. After careful selection in total 15 unique human proteins could be identified using mass spectrometry. Largest functional disease groups of proteins with significant differences in abundance were those related to hematological diseases, inflammatory and infectious diseases. Regarding physiological cellular functions the proteins altered in our study belonged to functional groups of protein synthesis, cellular comprise and cellular maintenance. Review of the literature of the specific proteins revealed several proteins that potentially play an essential role in puumala virus replication or endothelial cell functioning. For instance proteins like, Heterogeneous nuclear ribonucleoprotein K, heat shock 27 kDa protein 1-, Heat shock protein 90 kDa beta member 1 and vimentin- are proven to be essential, or important, host factors for the replication of certain viruses.

To summarize, we have identified a significant number of altered proteins in primary cultured endothelial cells during PUUV and *Leptospira interrogans* serovar Copenhageni infection. Several of these proteins are important factors in cell maintenance and functioning while also some of them have been proven to play an essential role in the replication of specific viruses. Future studies using knock down and elimination experiments will increase the understanding behind the mechanism of both diseases. Eventually a continued combined proteome approach could lead to new tailor made therapeutic targets.

References

Adler, B. and de la Pena, M.A., 2010. Leptospira and leptospirosis. Vet. Microbiol. 140: 287-296.

Bharti, A.R., Nally, J.E., Ricaldi, J.N., Matthias, M.A., Diaz, M.M., Lovett, M.A., Levett, P.N., Gilman, R.H., Willig, M.R., Gotuzzo, E. and Vinetz, J.M., 2003. Leptospirosis: a zoonotic disease of global importance. Lancet Infect. Dis. 3: 757-771.

Chierakul, W., Tientadakul, P., Suputtamongkol, Y., Wuthiekanun, V., Phimda, K., Limpaiboon, R., Opartkiattikul, N., White, N.J., Peacock, S.J. and Day, N.P., 2008. Activation of the coagulation cascade in patients with leptospirosis. Clin. Infect. Dis. 46: 254-260.

Ellis, W.A., 1994. Leptospirosis as a cause of reproductive failure. Vet. Clin. North Am. Food Anim Pract. 10: 463-478.

Gavrilovskaya, I.N., Gorbunova, E.E. and Mackow, E.R., 2010. Pathogenic hantaviruses direct the adherence of quiescent platelets to infected endothelial cells. J. Virol. 84: 4832-4839.

Gavrilovskaya, I.N., Peresleni, T., Geimonen, E. and Mackow, E.R., 2002. Pathogenic hantaviruses selectively inhibit beta3 integrin directed endothelial cell migration. Arch Virol. 147: 1913-1931.

Goeijenbier, M., Hartskeerl, R.A., Reimerink, J., Verner-Carlsson, J., Wagenaar, J.F., Goris, M.G., Martina, B.E., Lundkvist, A., Koopmans, M.P., Osterhaus, A.D., van Gorp, E.C. and Reusken, C.B., 14 A.D. The Hanta Hunting Study: Underdiagnosis of puumala hantavirus infections in symptomatic non-travelling leptospirosis-suspected patients in the Netherlands, in 2010 and April to November 2011. In: Anonymous, Country.

Goeijenbier, M., Wagenaar, J., Goris, M., Martina, B., Henttonen, H., Vaheri, A., Reusken, C., Hartskeerl, R., Osterhaus, A. and Van, G.E., 2013. Rodent-borne hemorrhagic fevers: under-recognized, widely spread and preventable – epidemiology, diagnostics and treatment. Crit Rev. Microbiol. 39: 26-42.

Gomez, R.M., Vieira, M.L., Schattner, M., Malaver, E., Watanabe, M.M., Barbosa, A.S., Abreu, P.A., de Morais, Z.M., Cifuente, J.O., Atzingen, M.V., Oliveira, T.R., Vasconcellos, S.A. and Nascimento, A.L., 2008. Putative outer membrane proteins of Leptospira interrogans stimulate human umbilical vein endothelial cells (HUVECS) and express during infection. Microb. Pathog. 45: 315-322.

Hamond, C., Pinna, A., Martins, G. and Lilenbaum, W., 2013. The role of leptospirosis in reproductive disorders in horses. Trop. Anim Health Prod.

Mackow, E.R. and Gavrilovskaya, I.N., 2009. Hantavirus regulation of endothelial cell functions. Thromb. Haemost. 102: 1030-1041.

Martinez-Lopez, D.G., Fahey, M. and Coburn, J., 2010. Responses of human endothelial cells to pathogenic and non-pathogenic Leptospira species. PLoS. Negl. Trop. Dis. 4: e918.

Taylor, S.L., Wahl-Jensen, V., Copeland, A.M., Jahrling, P.B. and Schmaljohn, C.S., 2013. Endothelial cell permeability during hantavirus infection involves factor XII-dependent increased activation of the kallikrein-kinin system. PLoS. Pathog. 9: e1003470.

Vieira, M.L., D'Atri, L.P., Schattner, M., Habarta, A.M., Barbosa, A.S., de Morais, Z.M., Vasconcellos, S.A., Abreu, P.A., Gomez, R.M. and Nascimento, A.L., 2007. A novel leptospiral protein increases ICAM-1 and E-selectin expression in human umbilical vein endothelial cells. FEMS Microbiol. Lett. 276: 172-180.

Wagenaar, J.F., Goris, M.G., Partiningrum, D.L., Isbandrio, B., Hartskeerl, R.A., Brandjes, D.P., Meijers, J.C., Gasem, M.H. and van Gorp, E.C., 2010. Coagulation disorders in patients with severe leptospirosis are associated with severe bleeding and mortality. Trop. Med. Int. Health 15: 152-159.

Profile changes in salivary glyco-enriched fraction of pigs with inflammation

A.M. Gutiérrez[1*], I. Miller[2], M. Fuentes-Rubio[1], K. Hummel[3], K. Nöbauer[3], E. Razzazi-Fazeli[3] and J.J. Cerón[1]

[1]Department of Animal Medicine and Surgery, Regional Campus of International Excellence 'Campus Mare Nostrum', University of Murcia, 30100 Espinardo, Murcia, Spain; agmontes@um.es
[2]Department of Biomedical Sciences, University of Veterinary Medicine Vienna, Veterinaerplatz 1, 1210 Vienna, Austria
[3]VetCORE Facility for Research, University of Veterinary Medicine Vienna, Veterinaerplatz 1, 1210 Vienna, Austria

Introduction

Glycosylation, mainly N-linked and O-linked glycosylation, is considered to be the most common form of post-translational modification in saliva (Sondej *et al.* 2009). N-glycans play important roles in protein folding and stability and are also involved in glycoprotein function in many cellular processes such as immune responses, developmental programs and cell-cell interactions, making the diversity of N-glycoproteins a source of potential disease biomarkers (Xu *et al.*, 2014).

The vast structural heterogeneity of glycans is based on the expression level and activity of glycosidases and glycosyltransferases, which can be altered in pathological conditions (Axford, 2001). Consequently, altered glycan structures are often attached to the same protein backbone as a consequence of a pathophysiological processes occurring in the cell that produces the protein (Gornik *et al.*, 2008). These alterations can be very specific and studies on protein glycosylation offer a good basis for diagnosis and prognosis of many diseases.

The objective of this work was to study the protein glycosylation pattern of porcine saliva in an inflammatory condition in order to see if specific changes in glycan structures, which would be useful in disease detection, could be found in this body fluid.

Material and methods

Different analytical procedures have been used to profile the human salivary glycoproteome to date. Recently we have established a protocol to study glycosylation changes in saliva proteins in pigs by using phenyl boronic acid (PBA) matrix. PBA is known to interact with cis-diol structures present in particular types of carbohydrates (Zeng *et al.*, 2013). The optimized protocol used in our study includes a complex sample preparation procedure in which several dialysis and incubation steps are needed (Figure 1).

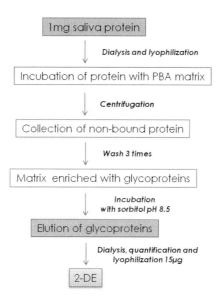

Figure 1. Optimized sample preparation procedure to study glycosylation changes in porcine saliva proteins by enrichment on phenyl boronic acid (PBA) matrix.

Saliva samples from a total of 10 pigs were obtained. All pigs were males, conventional Duroc × (Landrace × Large White) which came from the same farm in the Southeast of Spain. Five animals were randomly selected from a pen in which animals had no clinical signs of disease at the time of sampling. Afterwards other five animals suffering from rectal prolapse were selected. Saliva was collected from animals in the early morning at farm, by allowing the pigs to chew a sponge clipped to a thin metal rod during 1-2 minutes. Saliva was obtained by centrifugation of sponges inserted in specific tubes (Sarstedt, Nümbrecht, Germany) for 10 min at 3,000×g and stored at -80 °C until analysis.

2-DE was performed using 15 µg of eluted fractions in nonlinear immobilized pH-gradient strips (IPGs) pH 3-11 of 11 cm length under reducing and denaturing conditions in a Multiphor II electrophoresis chamber (GE Healthcare Life Sciences, Munich, Germany). Isoelectrofocusing conditions included a multistep protocol in which voltage was increased in gradient mode with a maximum of 3,500 V until 12 kVh were reached. For the second dimension, electrophoresis was carried out at 15 °C and 25 mA/gel for 4-5 hours in home-made 10-15% polyacrylamide gradient gels of 140×140×1.5 mm. Afterwards, gels were silver stained as previously reported (Gutiérrez *et al.*, 2011).

Silver stained 2-DE gels were digitalized using an ImageScanner II (GE Healthcare Life Sciences, Uppsala, Sweden) and evaluated by using a specific software (Image Master 2D Platinum 7.0, GE Healthcare Life Sciences, Uppsala, Sweden). Analysis included spot detection, landmarking and spot matching of protein patterns of all gels in the set. An unpaired t test was used to evaluate spot concentration differences between classes (healthy controls and prolapsed pigs) with a statistical program (GraphPad Prism 5, GraphPad software Inc.). The level of significance was set at $P<0.05$.

Results and discussion

For both groups of animals, similar ratios of proteins were bound to the PBA resin: 8.35% (SD=1.74) of whole saliva protein content in healthy pigs and 6.58% (SD=3.34) in prolapsed animals. Several statistically significant differences were observed when the glyco-patterns of healthy and prolapsed pigs were compared ($P<0.05$) (Figure 2).

The area in the box of Figure 2 contains the region with most changes noticed. Spots in this region and specifically this particular spot pattern ('vertical spot chains') have not been reported in literature in porcine saliva, neither in healthy animals nor in diseased pigs (Gutiérrez *et al.* 2011, 2013). A possible explanation could be changes in post-translational modifications, particularly glycosylation, or modifications due to the sample pretreatment/resin interaction, preferentially in the diseased animals, that produce a variation in the original pI. Further analysis concerning protein identification and type of modification is on the way, in order to clarify the origin of those spots, the reason for their altered mobility and the possible link with the inflammatory condition of those animals.

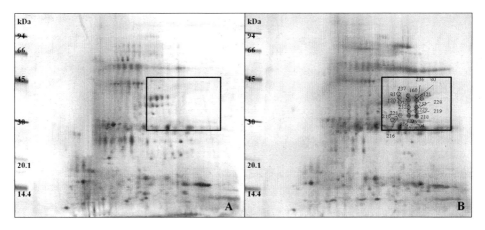

Figure 2. Images of salivary glyco-enriched patterns, obtained by two-dimensional electrophoresis, on samples from a control pig (A) and a pig suffering from rectal prolapse (B). Encircled and numbered spots were of statistically significantly higher concentration in pigs suffering from rectal prolapse.

References

Axford, J., 2001. The impact of glycobiology on medicine. Trends in immunology 22, 237-239.

Gornik, O. And Lauc, G., 2008. Glycosylation of serum proteins in inflammatory diseases. Disease Markers 25, 267-278.

Gutiérrez, A.M. Miller, I. Hummel, K. Nöbauer, K. Martínez-Subiela, S. Razzazi-Fazeli, E. Gemeiner, M. Cerón, J.J., 2011. Proteomic analysis of porcine saliva. Veterinary Journal 187, 356-362.

Gutiérrez, A.M. Nöbauer, K. Soler, L. Razzazi-Fazeli, E. Gemeiner, M. Cerón, J.J. Miller, I., 2013. Detection of potential markers for systemic disease in saliva of pigs by proteomics: a pilot study. Veterinary Immunology and Immunopathology 151, 73-82.

Sondej, M. Denny, P.A. Xie, Y. Ramachandran, P. Si, Y. Takashima, J, Shi, W. Wong, D.T. Loo, J.A. and Denny, P.C., 2009. Glycoprofiling of the Human Salivary Proteome. Clinical Proteomics 5, 52-68.

Xu, Y. Bailey, U. Punyadeera, C. and Schulz, B.L., 2014. Identification of salivary N-glycoproteins and measurement of glycosylation site occupancy by boronate glycoprotein enrichment and liquid chromatography/electrospray ionization tandem mass spectrometry. Rapid Communications in Mass Spectrometry 28, 471-482.

Zeng, Z. Wang, Y. Gui, X. Wang L. Lu, N., 2013. On-plate glycoproteins/glycopeptides selective enrichment and purification based on surface pattern for direct MALDI MS analysis. Analyst 138, 3032-3037.

A proteomics study on colostrum and milk proteins of the two major small ruminant dairy breeds from the Canary Islands on a bovine comparison perspective

Lorenzo E. Hernández-Castellano[1,2], André M. Almeida[3], Sébastien Planchon[4], Jenny Renaut[4], Anastasio Argüello[1] and Noemí Castro[1]*
[1]*Department of Animal Science, Universidad de Las Palmas de Gran Canaria, Arucas, Gran Canaria, Spain*
[2]*Veterinary Physiology, Vetsuisse Faculty, University of Bern, Bern, Switzerland;*
lorenzo.hernandez@vetsuisse.unibe.ch
[3]*Instituto de Investigação Científica Tropical, Lisboa, Portugal*
[4]*Centre de Reserche Public – Gabriel Lippmann, Dept. of Environment and Agrobiotechnologies, Luxembourg*

Introduction

Colostrum is the first secretion form the mammary gland after parturition and it starts changing after birth becoming mature milk (Hernández-Castellano *et al.*, 2014a; Lérias *et al.*, 2014). As newborn ruminants are considered agammaglobulinemic (calves) or hypo-gammaglobulinemic (lambs and kids) at birth (Castro *et al.*, 2009; Hernández-Castellano *et al.*, 2014b), colostrum feeding is very important to provide protection against infections (Kramer *et al.*, 2001). After the colostrum intake period, new-born ruminants must be fed with milk, however, there is an increasing number of high production dairy farms (Lérias *et al.*, 2013), such as the large-scale small ruminant dairy farms in the Canary Islands (Spain), where artificial rearing is chosen in order to increase the amount of milk available for processing (Napolitano *et al.*, 2008). This alternative feeding source has to be carefully selected, because many factors can modify the final components concentration of the ruminant milk, such as breed (Torres *et al.*, 2013) or milking frequency (Hernández-Castellano *et al.*, 2011), affecting the final performance and even the survival of the young ruminant. In this work we analyzed colostrum and milk differences between the two most economically relevant dairy ruminant species breeds from the Canary Islands: Canarian dairy sheep and Majorera goat. In order to obtain a comparison term, we have also contrasted our results to those of colostrum and milk of the most found ruminant species and breed across the globe: Holstein Friesian dairy cattle.

Material and methods

Samples collection and treatment for analysis

Six Holstein-Frisian cows, six Canarian dairy sheep and six Majorera goats in their second lactation were used in this experiment. Animals were fed following recommendations of the *Institut National de la Recherche Agronomique* (INRA, 2007). The experiment took place at the experimental farm of the Veterinary Faculty of the Universidad de Las Palmas de Gran Canaria (Spain) in spring.

Colostrum samples (10 ml) were taken immediately after partum in the three studied species. Milk samples (10 ml) were taken 20 days after partum in the ascending phase of lactation curve for the three species. Colostrum and milk samples were centrifuged following the procedure described by Boehmer *et al.* (2008) and the translucent supernatant (whey fraction) was collected and stored at -80 °C.

Two-dimensional gel electrophoresis, image analysis and protein identification

Colostrum and milk whey fractions were desalted with ReadyPrep 2D-Clean-up kit (Biorad, Hercules, CA, USA). Protein concentration was determined using the Quick Start Protein Assay kit (Biorad, Hercules, CA, USA). Approximately 400 µg of each whey protein sample was diluted in rehydration buffer (8 M urea, 2% (w/v) CHAPS, 50 mM dithiothreitol, 0.2% (w/v) BioLyte 3/10 ampholyte and 0.002% (w/v) Bromophenol Blue to a final sample volume of 300 µl. After dilution in rehydration buffer, samples were applied to 17-cm pH 3-10 nonlinear immobilized pH gradient strips (Biorad, Hercules, CA, USA) and focused in a Bio-Rad Protean IEF Cell for 20 h as described by Boehmer *et al.* (2008). Subsequently, strips were equilibrated as described by Hernández-Castellano *et al.* (2014a). Second dimension was conducted after equilibration using 12.5% polyacrylamide gels on a Protean II xi Cell electrophoresis system (Biorad, Hercules, CA, USA), using the running conditions as recommended by the manufacturer (1 W/gel for 1 h and 2 W/gel for 14-16 h at 12 °C). Each gel was stained using Coomassie Brillant Blue G-250 as previously described by Almeida *et al.* (2010) and scanned with a Gel Doc XR system (Biorad, Hercules, CA, USA). In order to detect differentially expressed proteins, gels were analyzed using Progenesis SameSpots software (Nonlinear Dynamics, Newcastle upon Tyne, UK). Spots with $P<0.05$ and a fold intensity higher than 1.4 were considered to have significantly different expression levels. Then, protein spots of interest were excised from the gel with a sterile 1,000 µl pipette tip or a sterile stainless scalpel blade for individual in-gel digestion using trypsin as described by Hernández-Castellano *et al.* (2014a). Peptide mass determinations were carried out according to Printz *et al.* (2013) using the 5800 Proteomics Analyzer (ABsciex) in reflectron mode for both peptide mass fingerprint and MS/MS. Calibration was performed with the peptide mass calibration kit for 4700 (ABsciex). Protein identification was done by searching the MS and MS/MS data against NCBI database in the Other Mammalia taxonomy (434586 sequences), using an in house MASCOT 2.3 server (www.matrixscience.com). Two trypsin missed cleavages, carbamidomethylation of cysteine as fixed modification as well as four dynamic modifications (methionine and tryptophan oxidation, tryptophan dioxydation and tryptophan to kynurenin) were allowed. Mass accuracy was set to 100 ppm for parent ions and 0.5 Da for MS/MS fragments. Homology identification was retained with probability set at 95%. All identifications were confirmed manually.

Results and discussion

In order to investigate different expressed proteins in colostrum and milk from cow, sheep and goat we used a Proteomics approach based on 2-D gel electrophoresis and a MALDI-TOF-TOF for protein identification. To the best of our knowledge, it is the first time that the colostrum and milk differences between these three species and breeds are compared using a proteomics approach. The differences in colostrum and milk from studied species can be observed in Table 1.

Table 1. Differential proteins identified in colostrum and milk from the three ruminant species (cow, sheep and goat).

Spot reference	Fold	Average normalized volumes			Protein name	Accession number
		Goat	Sheep	Cow		
Colostrum						
380	3.2	8.254e+006	1.337e+007	2.655e+007	Ig heavy chain C region	gi|109029
351	2.8	1.484e+007	3.529e+007	1.239e+007	albumin precursor	gi|193085052
338	3.3	1.007e+007	1.370e+007	4.195e+006	albumin precursor	gi|193085052
387	3.7	1.088e+007	8.348e+006	3.047e+007	immunoglobulin gamma 2 heavy chain constant region	gi|147744654
383	2.6	1.082e+007	1.537e+007	2.782e+007	Ig heavy chain C region	gi|109029
346	3.4	7.453e+006	1.303e+007	3.863e+006	serum albumin precursor	gi|57164373
391	2.7	1.610e+007	1.281e+007	3.425e+007	immunoglobulin gamma 2 heavy chain constant region	gi|147744654
390	2.0	5.129e+007	2.999e+007	5.946e+007	Ig heavy chain C region	gi|109029
275	2.8	1.249e+006	8.036e+005	2.216e+006	Serotransferrin	gi|2501351
384	2.0	1.601e+007	1.955e+007	3.138e+007	Ig heavy chain C region	gi|109029
385	1.7	2.540e+007	2.681e+007	4.310e+007	Ig gamma-1 chain	gi|346578
236	1.9	4.003e+006	7.109e+006	3.739e+006	polymeric immunoglobulin receptor isoform 1	gi|426239425
270	1.9	1.362e+006	1.289e+006	2.436e+006	serotransferrin precursor	gi|296490958
253	2.9	6.280e+006	1.109e+007	1.814e+007	immunoglobulin mu heavy chain constant region	gi|162424563
392	1.6	3.720e+007	3.645e+007	5.693e+007	Ig heavy chain C region	gi|109029
243	2.6	6.707e+006	9.927e+006	1.747e+007	immunoglobulin mu heavy chain constant region	gi|162424563
232	1.9	6.008e+006	1.146e+007	9.045e+006	polymeric immunoglobulin receptor isoform 1	gi|426239425
393	1.5	3.393e+007	2.206e+007	3.317e+007	immunoglobulin gamma 1 heavy chain constant region	gi|91982959

▶▶

Table 1. Continued.

Spot reference	Fold	Average normalized volumes			Protein name	Accession number
		Goat	Sheep	Cow		
Milk						
283	3.9	1.311e+006	1.398e+006	5.058e+006	Chain A, Crystal Struc. LPO At 2.4a Resolution	gi|158430634
462	3.4	1.013e+007	1.532e+007	4.570e+006	Ig heavy chain C region	gi|109029
359	2.4	1.019e+007	1.556e+007	6.576e+006	Albumin precursor	gi|193085052
461	3.1	1.019e+007	1.511e+007	4.915e+006	Ig heavy chain C region	gi|109029
229	2.6	3.393e+006	4.319e+006	1.657e+006	polymeric immunoglobulin receptor isoform 1	gi|426239425
456	1.9	6.412e+006	1.194e+007	6.324e+006	Ig heavy chain C region	gi|109029
464	2.4	7.028e+006	1.149e+007	4.700e+006	Ig heavy chain C region	gi|109029
445	2.6	1.124e+006	2.919e+006	1.358e+006	Ig heavy chain C region	gi|109029
448	2.2	8.415e+006	1.882e+007	9.842e+006	Ig heavy chain C region	gi|109029
288	3.1	5.487e+005	5.411e+005	1.665e+006	Chain A, Crystal Struc. LPO At 2.4a Resolution	gi|158430634
460	2.6	7.060e+006	9.220e+006	3.507e+006	Ig heavy chain C region	gi|109029
468	2.3	9.510e+006	1.472e+007	6.279e+006	Ig heavy chain C region	gi|109029
454	2.1	5.013e+006	1.033e+007	5.423e+006	Ig heavy chain C region	gi|109029
442	2.2	2.737e+006	6.108e+006	3.319e+006	Ig heavy chain C region	gi|109029

Even procedure described by Boehmer *et al.* (2008) reduced the casein and immunoglobulin (Ig) concentration in raw colostrum and milk samples, it was still found an important presence of Ig in both whey fractions, being also found some caseins in colostrum whey. This fact could mask the presence of low abundance proteins (LAP) by Ig's and caseins. Nevertheless, our results showed 8 proteins in colostrum and 4 in milk a higher presence of albumin precursor in goat and sheep colostrum and milk than in cow colostrum and milk. On contrast, a higher presence of serotransferrin precursor in bovine colostrum than in the respective from goat and sheep was detected. Additionally, lactoperoxidase was higher detected in cow whey fractions than either goat or sheep. Additionally, different Ig's composition was observed among different studied species not only in colostrum but also in milk. The study provides a solid foundation for future studies that could in fact lead to the establishment of important proteins to consider when developing an alternative colostrum or milk source used in artificial reared ruminants, although different depletion techniques must be performed in order to increase the relative concentration of LAP.

Conclusions

The results of this work described for the first time differences among dairy ruminant species (cow, sheep and goat) in 8 colostrum and 4 milk proteins. Proteins from both fluids (colostrum and milk) were relatively similar in goats and sheep. They were however different when compared to bovine samples. Understanding the differences in colostrum and milk proteins among dairy ruminant species is of high relevance aspects such as the heterologous passive immunity (i.e. immunoglobulins obtained from one species and utilized for passive immunity in another species). Further studies will be necessary in order to increase the general knowledge about differences between these three dairy ruminant species.

Acknowledgements

Authors acknowledge financial support from the Formación del Profesorado Universitario (FPU) program (Ministry of Education, Madrid, Spain) as well as the program Ciência 2007 from *Fundação para a Ciência e a Tecnologia* (Lisbon, Portugal).

References

Almeida, A.M., Campos, A., Francisco, R., van Harten, S., Cardoso, L.A. and Coelho, A.V., 2010. Proteomic investigation of the effects of weight loss in the gastrocnemius muscle of wild and NZW rabbits via 2D-electrophoresis and MALDI-TOF MS. Animal Genetics 41: 260-272.

Boehmer, J.L., Bannerman, D.D., Shefcheck, K. and Ward, J.L., 2008. Proteomic Analysis of Differentially Expressed Proteins in Bovine Milk During Experimentally Induced Escherichia coli Mastitis. Journal of Dairy Science 91: 4206-4218.

Castro, N., Capote, J., Morales-Delanuez, A., Rodriguez, C. and Arguello, A., 2009. Effects of newborn characteristics and length of colostrum feeding period on passive immune transfer in goat kids. J Dairy Sci 92: 1616-1619.

Hernández-Castellano, L., Almeida, A., Ventosa, M., Coelho, A., Castro, N. and Arguello, A., 2014a. The effect of colostrum intake on blood plasma proteome profile in newborn lambs: low abundance proteins. Bmc Veterinary Research 10: 85.

Hernández-Castellano, L.E., Almeida, A.M., Castro, N. and Argüello, A., 2014b. The colostrum proteome, ruminant nutrition and immunity: a review. Current protein and peptide science 15: 64-74.

Hernández-Castellano, L.E., Torres, A., Alavoine, A., Ruiz-Diaz, M.D., Arguello, A., Capote, J. and Castro, N., 2011. Effect of milking frequency on milk immunoglobulin concentration (IgG, IgM and IgA) and chitotriosidase activity in Majorera goats. Small Ruminant Research 98: 70-72.

Kramer, M.S., Chalmers, B., Hodnett, E.D., Sevkovskaya, Z., Dzikovich, I., Shapiro, S., Collet, J.P., Vanilovich, I., Mezen, I., Ducruet, T., Shishko, G., Zubovich, V., Mknuik, D., Gluchanina, E., Dombrovskiy, V., Ustinovitch, A., Kot, T., Bogdanovich, N., Ovchinikova, L., Helsing, E. and Grp, P.S., 2001. Promotion of breastfeeding intervention trial (PROBIT) – A randomized trial in the Republic of Belarus. Jama-Journal of the American Medical Association 285: 413-420.

Lérias, J., Hernández-Castellano, L., Morales-delaNuez, A., Araújo, S., Castro, N., Argüello, A., Capote, J. and Almeida, A., 2013. Body live weight and milk production parameters in the Majorera and Palmera goat breeds from the Canary Islands: influence of weight loss. Trop Anim Health Prod 45: 1731-1736.

Lérias, J.R., Hernández-Castellano, L.E., Suárez-Trujillo, A., Castro, N., Pourlis, A. and Almeida, A.M., 2014. The mammary gland in small ruminants: major morphological and functional events underlying milk production – a review. Journal of Dairy Research 81: 304-318.

Napolitano, F., Pacelli, C., Girolami, A. and Braghieri, A., 2008. Effect of information about animal welfare on consumer willingness to pay for yogurt. Journal of Dairy Science 91: 910-917.

Printz, B., Sergeant, K., Guignard, C., Renaut, J. and Hausman, J.F., 2013. Physiological and proteome study of sunflowers exposed to a polymetallic constraint. Proteomics 13: 1993-2015.

Torres, A., Castro, N., Hernandez-Castellano, L.E., Arguello, A. and Capote, J., 2013. Effects of milking frequency on udder morphology, milk partitioning, and milk quality in 3 dairy goat breeds. Journal of Dairy Science 96: 1071-1074.

A 2DE map of the urine proteome in the cat: effect of Chronic Kidney Disease

Enea Ferlizza[1], Alexandre Campos[2], Aurora Cuoghi[3], Elisa Bellei[3], Emanuela Monari[3], Francesco Dondi[1], André M. Almeida[4,5] and Gloria Isani[1]*

[1]*Dept. of Veterinary Medical Sciences, University of Bologna, via Tolara di sopra 50, 40064 Ozzano, Bologna, Italy; gloria.isani@unibo.it*
[2]*Interdisciplinary Centre of Marine and Environmental Research, University of Porto, rua dos Bragas 289, 4050-123 Porto, Portugal*
[3]*Department of Diagnostic, Clinical and Public Health Medicine, University of Modena and Reggio Emilia, via del Pozzo 71, 41124 Modena, Italy*
[4]*CIISA/IICT – CVZ, faculdade de Medicina Veterinária, ULisboa, Av. Univ. Técnica, 1300-477 Lisboa, Portugal*
[5]*IBET & ITQB/UNL, Oeiras, Portugal*

Introduction

Urine is considered an ideal source of clinical biomarkers as it can be obtained noninvasively, repeatedly and in adequate amounts. In veterinary medicine, the application of proteomics techniques is still very limited. The aim of our work was to produce a preliminary map of the urine proteome of the healthy cats (*Felis catus*) and to compare it with the proteome of cats affected by chronic kidney disease (CKD). For that we have used an approach based on two-dimensional electrophoresis and protein identification using mass spectrometry (MS).

Materials and methods

Urine samples were collected by cystocentesis from 4 healthy and 4 cats affected by CKD and analyzed by 2DE. The first dimension was performed by isoelectric focusing on 17 cm long IPG strips (pH 3-10); the second dimension was performed on 10% SDS-PAGE and stained with colloidal Coomassie. Spots were excised from the gel, reduced, alkylated and digested with trypsin and identified using ESI-Q-TOF MS (Campos *et al.*, 2013).

Results and discussions

2DE allowed the separation of 66 spots in the urine proteome of healthy and CKD cats. Eighteen spots were overrepresented in CKD and nine spots were underrepresented. The 27 differentially expressed spots and the nine most abundant common spots were excised from the gels for MS identification (Figure 1; Table 1).

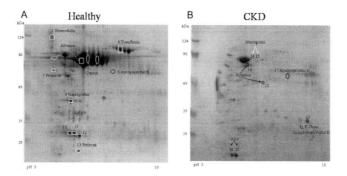

Figure 1. 2DE of the urine proteome in healthy (A) and CKD (B) cats. White circles: spots with significantly greater intensity in healthy than in CKD; black circles: spots with significantly greater intensity in CKD; white rectangles: common spots without significant differences.

Preliminary 2D map

21 spots yielded significant results by MS, producing a preliminary feline urine map, including 13 proteins that may be functionally classified as transport (38%), immune and cellular response (38%), and cellular communication and growth (15%). The most abundant protein was cauxin, a serine esterase produced by healthy tubular cells, specifically excreted in urine of cats and probably involved in the synthesis of felinine pheromone (Miyazaki *et al.*, 2007). The transport proteins, albumin, transferrin, haemopexin and haptoglobin all derive from plasma and have been identified as common components of urine also from healthy humans (Candiano *et al.*, 2010). Among the proteins involved in immune and cellular defence response, we identified IgK light chain, protein AMBP and uromodulin. Differently from dogs (Brandt *et al.*, 2014) and humans (Lhotta, 2010), uromodulin is not the most abundant urine-specific protein in cats. The remaining proteins, perlecan and fetuin-A, are involved in cell communication and growth. In particular, perlecan, a negatively charged proteoglycan of the glomerular filtration barrier, has also been identified in dog urine (Nabity *et al.*, 2011).

Effect of CKD

Regarding the effect of CKD on the urine proteome, seven differentially represented proteins have been identified These proteins can be indicative of tubular dysfunction when not reabsorbed (e.g. RBP) or not secreted (e.g. uromodulin and cauxin) and could be studied as putative biomarkers of nephropathy. Among the overrepresented proteins, retinol binding protein (RBP) is a 22 kDa protein freely filtered by the glomerulus and reabsorbed by the tubules. The appearance of RBP in urine is a marker of impaired tubular function in humans (Pallet *et al.*, 2014) and it has also been reported in dogs (Nabity *et al.*, 2011) and cats (van Hoek *et al.*, 2008). Interesting underrepresented proteins were uromodulin and cauxin. Uromodulin is a glycoprotein produced by healthy tubular cells and its disappearance has been already proved in humans affected by CKD (Lhotta, 2010) and

Table 1. Proteins identified in cat urine by mass spectrometry.

Spot	Entry name[1]	Protein full name	MW (kDa)	pI	Score[2]	Pept.[3]	Sign. seq.[4]
1	UROM_CANFA	Uromodulin	72.9	4.94	130	36	3
2	ALBU_FELCA	Serum albumin	70.6	5.46	2,383	196	28
3	ALBU_FELCA	Serum albumin	70.6	5.46	2,133	208	29
4	EST5A_FELCA	Carboxylesterase 5A	60.9	5.58	524	66	10
5	EST5A_FELCA	Carboxylesterase 5A	60.9	5.58	447	89	10
6	TRFE_PIG	Serotransferrin	78.9	6.93	114	31	5
7	FETUA_HUMAN	Fetuin-A	40.1	5.43	141	34	4
8	APOH_CANFA	Apolipoprotein H	39.7	8.51	162	21	4
9	HPT_BOVIN	Haptoglobin	45.6	7.83	72	6	2
10	AMBP_BOVIN	Protein AMBP	40.1	7.81	141	5	1
11	AMBP_BOVIN	Protein AMBP	40.1	7.81	150	6	1
12	AMBP_BOVIN	Protein AMBP	40.1	7.81	274	11	1
13	PGBM_HUMAN	Perlecan	479.3	6.06	134	19	2
14	HEMO_PONAB	Hemopexin	52.3	6.44	73	25	1
15	HEMO_PONAB	Hemopexin	52.3	6.44	97	25	1
16	ALBU_FELCA	Serum albumin	70.6	5.46	1,585	187	25
17	APOH_CANFA	Apolipoprotein H	39.7	8.51	119	16	4
18	ALBU_FELCA	Serum albumin	70.6	5.46	69	10	3
19	KV1_CANFA	Ig kappa chain V region GOM	12.1	6.41	111	4	2
	CFAD_PIG	Complement factor D	28.3	6.59	54	9	2
20	RET4_HORSE	Retinol-binding protein 4	23.3	5.28	60	4	1
21	RET4_HUMAN	Retinol-binding protein 4	23.3	5.76	167	27	3

[1] Protein entry name from UniProt knowledge database.

[2] The highest scores obtained with Mascot search engine.

[3] Peptides: total number of peptides matching the identified proteins.

[4] Significant Sequences: total number of significant sequences matching the identified proteins.

could be applied also in cats. Regarding cauxin, according to Miyazaki *et al.*, (2007) this protein could be a promising biomarker for the determination of tubular damage in CKD cats.

Conclusions

2DE was essential in fractionation of the complex urine proteome in cats, producing a preliminary map that included 13 proteins. In particular, uromodulin, cauxin and perlecan, specifically secreted in urine, could help in the evaluation of renal function. Seven proteins were differentially represented in CKD cats suggesting their use as a putative biomarker of nephropathy in cats and possibly in other veterinary species.

References

Brandt, L.E., Ehrhart, E.J., Scherman, H., Olver, C.S., Bohn, A., Prenni, J.E., 2014. Characterization of the canine urinary proteome. Veterinary Clinical Pathology 43, 193-205.

Campos, A., Puerto, M., Prieto, A., Cameán, A., Almeida, A.M., Coelho, A.V, Vasconcelos, V., 2013. Protein extraction and two-dimensional gel electrophoresis of proteins in the marine mussel Mytilus galloprovincialis: an important tool for protein expression studies, food quality and safety assessment. Journal of the Science of Food and Agricolture 93, 1779-1787.

Candiano, G., Santucci, L., Petretto, A., Bruschi, M., Dimuccio, V., Urbani, A., Bagnasco, S., Ghiggeri, G.M., 2010. 2D-electrophoresis and the urine proteome map: where do we stand? Journal of Proteomics 73, 829-844.

Lhotta, K., 2010. Uromodulin and chronic kidney disease. Kidney and Blood Pressure Research 33, 393-398.

Miyazaki, M., Soeta, S., Yamagishi, N., Taira, H., Suzuki, A., Yamashita, T., 2007. Tubulointerstitial nephritis causes decreased renal expression and urinary excretion of cauxin, a major urinary protein of the domestic cat. Research in Veterinary Science 82, 76-79.

Nabity, M.B., Lees, G.E., Dangott, L.J., Cianciolo, R., Suchodolski, J.S., Steiner, J.M., 2011. Proteomic analysis of urine from male dogs during early stages of tubulointerstitial injury in a canine model of progressive glomerular disease. Veterinary Clinical Pathology 40, 222-236.

Pallet, N., Chauvet, S., Chassé, J.-F., Vincent, M., Avillach, P., Levi, C., Meas-Yedid, V., Olivo-Marin, J.-C., Nga-Matsogo, D., Beaune, P., Thervet, E., Karras, A., 2014. Urinary retinol binding protein is a marker of the extent of interstitial kidney fibrosis. PLoS One 9, e84708.

Van Hoek, I., Daminet, S., Notebaert, S., Janssens, I., Meyer, E., 2008. Immunoassay of urinary retinol binding protein as a putative renal marker in cats. Journal of Immunological Methods 329, 208-213.

Serum proteomic analysis of zoonotic-related abortion in cows

Matko Kardum[1]*, Cristian Piras[2], Alessio Soggiu[2], Viviana Greco[3], Paola Roncada[2,4], Marko Samardžija[5], Silvio Špičić[6], Dražen Đuričić[7], Nina Poljičak Milas[1] and Romana Turk[1]

[1]*Department of Pathophysiology, Faculty of Veterinary Medicine, University of Zagreb, Croatia;*
matko.kardum@gmail.com
[2]*Department of Veterinary Science and Public Health, University of Milan, Milan, Italy*
[3]*Fondazione Santa Lucia – IRCCS, Rome, Italy*
[4]*Istituto Sperimentale Italiano L. Spallanzani, Milano, Laboratory of Microbial Proteomics, Italy*
[5]*Department of Reproduction and Clinic for Obstetrics, Faculty of Veterinary Medicine, University of Zagreb, Croatia*
[6]*Laboratory for Bacterial Zoonoses and Molecular Diagnostic of Bacterial Diseases, Croatian Veterinary Institute, Zagreb, Croatia*
[7]*Veterinary Practice Djurdjevac, Croatia*

Introduction

Abortion in cows represents a serious burden for dairy industry with a great economic impact. Most abortions are due to infectious diseases, several of them are zoonoses. The risk of infection with zoonotic pathogens is not only relevant for animal health and welfare but also for human health due to the shedding of pathogens into environment as well as contamination of animal products such as milk. Bacterial zoonoses such as leptospirosis, Q fever and brucellosis are important cattle infections associated with intrauterine death and reduced fertility [1]. Zoonoses are often inapparent or subclinical diseases and can be manifested only with abortion [2]. Intrauterine foetal death results from foetus infections with zoonotic pathogens due to the passage of pathogens through the placenta mostly during late pregnancy.

Leptospirosis is considered an emerging zoonosis not only in developing rural areas, but also in the urban areas of developed countries. It is caused by infection with pathogenic spirochetes of the genus *Leptospira* [3]. Q fever is caused by the bacterium *Coxiella burnetii* and domestic ruminants are the primary source of infection for humans [4].

Common mechanisms involved in the process of abortion still remain to be investigated since abortion often represents the only clinical appearance of disease.

The aim of this study is to explore differential proteome profile in serum of cows that aborted due to the subclinical infection with *Leptospira* and *Coxiella*. The main target is to uncover the pathophysiological molecular mechanisms involved in abortion caused by infectious diseases.

Materials and methods

The study was conducted on 15 dairy cows located on farms in Croatia. Cows were divided into three groups: Group I (control, n=5) consisted of healthy cows; Group II (cows infected with *Coxiella*, n=5) and Group III (cows infected with *Leptospira*, n=5) both comprised cows that have aborted. In case of Group II and Group III, bovine sera were collected immediately after abortion and serologically tested with the referent serological methods in order to detect the serological evidence of leptospirosis and Q fever. In case of Group I (control group) sera were taken approximately 100 days after delivery and also serologically tested on these pathogens with negative results. All serum samples were stored at -70 °C until analysis.

From each of 15 collected bovine sera chosen for the experiment 10 µl of sera was resuspended in 190 µl of 2D buffer (7 M urea, 2 M thiourea, 4% CHAPS). Protein assay was performed using BioRad Protein assay according to the manufacturer protocol. 1D- and 2D-electrophoresis were then carried out. After run, gels were stained with colloidal Coomassie Blue G250 solution and de-stained in MiliQ water. The stained gels were scanned using ImageScanner III and image analysis was performed by Progenesis SameSpots software (Nonlinear Dynamics) using 2D analysis module. After spot cutting from the gels and protein digestion with trypsin, the mass spectra were acquired with an Ultraflex III MALDI-TOF/TOF spectrometer (Bruker-Daltonics) and analyzed by MASCOT algorithm (www.matrixscience.com) against SwissProt_201407 database.

Results and discussion

Six proteins were found to be differentially expressed ($P \leq 0.05$) among the investigated groups. Among these, actin (spot 696, P63258), fructose-1,6-bisphosphatase 1 (spot 733, Q3SZB7) and transgelin-2 (spot 877, Q5E9F5) were found to be up-regulated in both leptospirosis and Q fever in comparison to the control. E3 ubiquitin-protein ligase KCMF1 (spot 655, Q1LZE1) and autophagy-related protein 13 (spot 594, Q08DY8) were found to be down-regulated both in leptospirosis and Q fever in comparison to the control. Gap junction alpha-5 protein (spot 747, Q0VCR2) was found to be down-regulated in Q fever but not in leptospirosis. This data and the whole 2D map are resumed in Figure 1 and 2. Accession numbers of proteins identified are reported in Figure 2.

All differentially expressed proteins were found to be up-regulated or down-regulated in bovine sera with Q fever and in leptospirosis in comparison to the control. This homology between the two pathologies might suggest that they share some common features that can cause abortion.

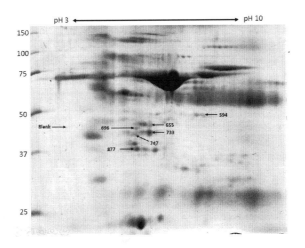

Figure 1. A representative 2DE of bovine sera indicating the differentially expressed proteins.

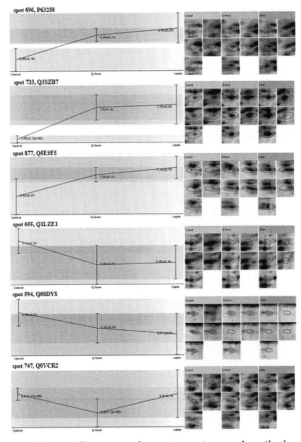

Figure 2. Details of the differentially expressed proteins as in text described.

Acknowledgements

This work was performed during a COST Short Term Scientific Mission (STSM) at the Instituto Sperimentale Spallanzani and the University of Milan.

References

[1] Barr, B., Anderson, M., 1993. Infectious diseases causing bovine abortion and fetal loss. The Veterinary Clinics of North America. Food Animal Practice 9, 343-368.

[2] Viana, K.F. and Zanini, M.S., 2012. Zoonotic abortion in herds: Etiology, prevention and control. In: Lorenzo-Morales, J. (ed.), Zoonosis, In Tech, Rijeka, Croatia, pp. 371-393.

[3] Bharti, A. R., Nally, J. E., Ricaldi, J. N., Matthias, M. A., *et al.*, 2003. Leptospirosis: a zoonotic disease of global importance. The Lancet. Infectious Diseases 3, 757-771.

[4] Guatteo, R., Seegers, H., Taurel, A.-F., Joly, A., Beaudeau, F., 2011. Prevalence of Coxiella burnetii infection in domestic ruminants: A critical review. Veterinary Microbiology 149, 1-16.

Comparison of quail and chicken broiler circulating IGF-1

Virge Karus[1]*, Avo Karus[1], Harald Tikk[2], Aleksander Lember[2] and Mati Roasto[3]
[1]Department of Chemistry, Institute of Veterinary Medicine and Animal Sciences, Eesti Maaülikool, Kreutzwaldi 62, Tartu 51014, Estonia; avo.karus@emu.ee
[2]Department of Small Farm Animal and Poultry Husbandry, Institute of Veterinary Medicine and Animal Sciences, Eesti Maaülikool, Kreutzwaldi 62, Tartu 51014, Estonia
[3]Department of Food Hygiene and Control, Institute of Veterinary Medicine and Animal Sciences, Eesti Maaülikool, Kreutzwaldi 62, Tartu 51014, Estonia

Introduction

Quail and chicken are highly important in regional agricultural poultry meat and egg production (Tikk *et al.*, 2009). Proteins are determined by exon sequences in animal genome. Also regulatory genes do have a great importance and therefore many studies are dealing with genome and QTL. Kayang *et al.* (2006) suggest that a wealth of information can be mined in chicken, to be used for genome analyses in quail. Similar study of Shiina *et al.* (2004) showed that the quail haplotype A spans 180 kb of genomic sequence, encoding a total of 41 genes compared with only 19 genes within the 92-kb chicken *Mhc*. Except for two gene families (*B30* and *tRNA*), both species have the same basic set of gene family members that were previously described in the chicken 'minimal essential' *Mhc*. Comparisons of quail and chicken suggested that the quail *Mhc* genes were duplicated after the separation of these two species from their common ancestor (Shiina *et al.*, 2004). Despite of these findings, full genome of quail is not sequenced yet. For comparison of these species we used IGF-1 (insulin-like growth factor-1) – a peptide believed to play an important role in the regulation of cellular growth and differentiation. IGF-1 is synthesized and secreted by many tissues and can act as endocrine hormone, which is transported by the circulation to distant sites of action, but it can also act locally by paracrine or autocrine mechanisms (Beccavin *et al.*, 2001; Giachetto *et al.*, 2004; Heck *et al.*, 2003). IGF-1 is also studied as a possible indicator for breeding in poultry (Beccavin *et al.*, 2001). IGF-1 has been shown to play significant role in chicken muscle development in all phases (Karus and Karus, 2003; Kocamis and Killefer, 2003; Saprõkina *et al.*, 2009). The aim of study was to examine IGF-1 gene-expression and circulating IGF-1 content in quail and chicken.

Material and methods

Trials were carried out according to Estonian Animal Protection Act (13.12.2000/RT I 2004). 21-day-old Estonian quail (*Coturnix coturnix*) and broilers were randomly assigned to feeding groups. The birds were fed starter diet until 21 days of age followed by finishing diet containing 4% rapeseed oil from day 21 to day 42. At the age of 42 days, blood samples were collected from 10 chickens (5 male and 5 female birds of the body weight similar to the sex's average) in each group. Tissue samples were taken immediately after slaughter (blood samples were taken from live birds before slaughtering), instantly frozen in dry ice and stored (about 2 hours) at -20 °C until use. Blood was collected from *V. jugularis* into disposable non-heparinized test tubes for IGF-1 testing and into EDTA-diNa tubes for mRNA studies. IGF-1 in blood serum was measured by RIA (DSL) kit. Blood serum was separated by centrifugation at 2,000×g for 10 min and was then frozen (-24 °C) for analysis. Whole blood samples were stabilized using RNA/DNA Stabilization Reagent for Blood/Bone Marrow (Roche Applied Science), and then frozen (-24 °C) for analysis. The LightCycler RNA Master SYBR Green I kit (Roche Applied Science) was used for hot start one-step RT-PCR in glass capillaries using the LightCycler Instruments and SYBR Green I dye as the detection format as described earlier (Karus *et al.*, 2007). GAPDH gene was used for IGF-1 mRNA relative quantification as housekeeping-gene. Data was analyzed using statistical program SYSTAT 11.0. All data of IGF-1 mRNA were tested for homogeneity of variance by the one-way ANOVA. When variances were comparable, the effect of nutritional state on IGF-1 mRNA synthesis was tested by T-test with unequal variances. After that Pearson's correlation coefficients were also determined. Differences at $P<0.05$ were regarded as significant.

Results and discussion

IGF-1 content in quail and hen broiler blood plasma did not differ in species or sex (Table 1).

It is important to mention that, in our experiments, we did not find any significant correlation between circulating IGF-1 and bird body weight. Despite of the relatively high variation in IGF-1 mRNA means (Karus *et al.*, 2012) we measured the gene-expression (relative to GAPDH) of IGF-1 also in blood. Clear downregulation of IGF-1 gene-expression in female birds was observed during finishing dietary period (days from 21 till 42). However, male birds showed tendency to upregulate it in the same period, but this change was statistically non-significant. The difference of IGF-1 mRNA content in leukocytes between female and male broilers at age of 3 weeks was detected ($P<0.05$). As described earlier, the IGF-1 level in plasma increases approximately six- to seven-fold from hatching to day 21 (Giachetto *et al.*, 2004). There was no clear correlation between the growth of the birds and hepatic IGF-1 mRNA expression and plasma IGF-1 levels. It has been shown, that in comparison of two parameters: IGF-1 protein content and IGF-1 relative gene-expression, there is clear advantage to measuring native free protein since the distribution of birds by mRNA level is significantly asymmetric, while native protein content is still close to normal distribution (Karus *et al.*, 2012). Burnside and Cogburn (1992) reported that the hepatic expression of IGF-1 mRNA in broiler chickens peaked at 28 days of age. Giachetto *et al.* (2004) and Yun *et al.* (2005) showed, that expression of IGF-1 mRNA in liver increased from hatch till seventh week, this result would be consistent with our recent observation of double increasing in IGF-1 mRNA

Table 1. IGF-1 content in quail and hen broiler blood plasma (µg/l) (mean±std).

Sex	Quail	Hen broiler
Female	2.93±1.70	2.24±1.60
Male	3.30±1.13	2.32±1.61

expression in liver at the nutrition period 22-42 days (*P*<0.1). Beccavin *et al.* (2001) found that high growth rate birds had a five-fold increase in IGF-I expression from the first to sixth week of life. Guernec *et al.* (2004) suggests that muscle IGF-1 mRNA levels are sensitive to the nutritional state of the chickens and may contribute to the effect of nutritional state on muscle development. McMurtry *et al.* (1997) has discussed that effect of nutritional status on muscle growth could be partly mediated by paracrine regulator, such as IGF-s. Most (approximately 80%) of circulating IGF-1 is synthesized in liver, and released into the bloodstream where it acts in endocrine way on other tissues. Our results do support the opinion, that endocrine action of IGF-1 is minimised and IGF-1 acts on paracrine manner. The specific mechanism why the clear downregulation of IGF-1 gene in 42 weeks old females will not result in lower circulating IGF-1 concentration remains still unclear and needs further investigations.

Acknowledgements

This work was supported by the pilot project P0091LATD04.

References

Beccavin, C., Chevalier, B., Cogburn, L., Simon, J. and Duclos, M., 2001. Insulin-like growth factors and body growth in chickens divergently selected for high or low growth rate. Journal of Endocrinology 168: 297-306.

Burnside, J. and Cogburn, L.A., 1992. Developmental expression of hepatic growth hormone receptor and insulin-like growth factor-I mRNA in the chicken. Molecular and Cellular Endocrinology 89: 91-96.

Giachetto, P.F., Riedel, E.C., Gabriel, J.E., Ferro, M.I.T., Di Mauro, S.M.Z., Macari, M. and Ferro, J.A., 2004. Hepatic mRNA expression and plasma levels of insulin-like growth factor-I (IGF-I) in broiler chickens selected for different growth rates. Genetics and Molecular Biology 27: 39-44.

Guernec, A., Chevalier, B. and Duclos, M.J., 2004. Nutrient supply enhances both IGF-I and MSTN mRNA levels in chicken skeletal muscle. Domestic Animal Endocrinology 26: 143-154.

Heck, A., Metayer, S., Onagbesan, O.M. and Williams, J., 2003. mRNA expression of components of the IGF system and of GH and insulin receptors in ovaries of broiler breeder hens fed ad libitum or restricted from 4 to 16 weeks of age. Domestic Animal Endocrinology 25: 287-294.

Karus, A. and Karus, V., 2003. IGF-1 and thyroid markers as prognostic tool for development. 15th IFCC-FESCC European Congress of Clinical Chemistry and Laboratory Medicine EUROMEDLAB 2003. Monduzzi Editore, Barcelona, pp. 513-516.

Karus, A., Saprõkina, Z., Tikk, A., Järv, P., Soidla, R., Lember, A., Kuusik, S., Karus, V., Kaldmäe, H., Roasto, M. and Rei, M., 2007. Effect of Dietary Linseed on Insulin-Like Growth Factor-1 and Tissue Fat Composition in Quails. Archiv für Geflügelkunde/European Poultry Science 71: 81-87.

Karus, A., Tikk, H., Lember, A., Karus, V. and Roasto, M., 2012. IGF-1 free circulating protein and its geneexpression in linseed-rich diet quail. In: P. Rodrigues, D. Eckersall and A. de Almeida (Eds.), Farm animal proteomics. Wageningen Academic Publishers, pp. 141-144.

Kayang, B., Fillon, V., Inoue-Murayama, M., Miwa, M., Leroux, S., Feve, K., Monvoisin, J.-L., Pitel, F., Vignoles, M., Mouilhayrat, C., Beaumont, C., Ito, S.i., Minvielle, F. and Vignal, A., 2006. Integrated maps in quail (Coturnix japonica) confirm the high degree of synteny conservation with chicken (Gallus gallus) despite 35 million years of divergence. BMC Genomics 7: 101.

Kocamis, H. and Killefer, J., 2003. Expression profiles of IGF-I, IGF-II, bFGF and TGF-beta 2 growth factors during chicken embryonic development. TURKISH JOURNAL OF VETERINARY & ANIMAL SCIENCES 27: 367-372.

McMurtry, J.P., Francis, G.L. and Upton, Z., 1997. Insulin-like growth factors in poultry. Domestic Animal Endocrinology 14: 199-229.

Saprōkina, Z., Karus, A., Kuusik, S., Tikk, H., Järv, P., Soidla, R., Lember, A., Kaldmäe, H., Karus, V. and Roasto, M., 2009. Effect of dietary linseed supplements on omega-3 PUFA content and on IGF-1 expression in broiler tissues. Agricultural and food science 18: 35-44.

Shiina, T., Shimizu, S., Hosomichi, K., Kohara, S., Watanabe, S., Hanzawa, K., Beck, S., Kulski, J.K. and Inoko, H., 2004. Comparative genomic analysis of two avian (quail and chicken) MHC regions. The Journal of Immunology 172: 6751-6763.

Tikk, H., Lember, A., Karus, A., Tikk, V. and M., P., 2009. Meat performance and meat chemical composition of quail broilers in Estonia. Journal of Agricultural Science XX: 47-59.

Yun, J., Seo, D., Kim, W. and Ko, Y., 2005. Expression and relationship of the insulin-like growth factor system with posthatch growth in the Korean Native Ogol chicken. Poultry science 84: 83-90.

A proteomic profile of uncomplicated and complicated babesiosis in dogs

Josipa Kuleš[1]*, Carlos de Torre[2], Renata Barić Rafaj[1], Jelena Selanec[3], Vladimir Mrljak[3] and Jose J. Ceron[4]

[1]Department of Chemistry and Biochemistry, Faculty of Veterinary Medicine, University of Zagreb, Croatia; jkules@vef.hr
[2]Proteomics Unit, Universitary Hospital Virgen de la Arrixaca, El Palmar – Murcia, Spain
[3]Internal Diseases Clinic, Faculty of Veterinary Medicine, University of Zagreb, Croatia
[4]Department of Animal Medicine and Surgery, Regional Campus of International Excellence 'Campus Mare Nostrum', University of Murcia, Murcia, Spain

Objectives

Canine babesiosis is a tick-borne disease that is caused by the haemoprotozoan parasites of the genus *Babesia* (Taboada and Merchant, 1991). The disease can be clinically classified into uncomplicated and complicated forms (Jacobson and Clark, 1994). Uncomplicated babesiosis has been suggested to be a consequence of anaemia resulting from haemolysis, whereas complicated canine babesiosis may be a consequence of the development of systemic inflammatory response syndrome (SIRS) and multiple organ dysfunction syndrome (MODS), both of which are cytokine-mediated phenomena (Matijatko *et al.*, 2012). The aim of this study was a comparative analysis of proteome in different clinical presentations of babesiosis, in order to identify potential proteins that can distinguish patients with SIRS or MODS from healthy dogs.

Material and methods

Sample collection and preparation

Blood samples were collected from 15 dogs of various breeds, sex (11 males and 4 females) and age (5 months to 10 years) with naturally occurring babesiosis caused by *B. canis canis*. According to SIRS and MODS criteria (Okano *et al.*, 2002; Welzl *et al.*, 2001) animals were divided as SIRS positive, SIRS negative (non-SIRS) and MODS positive (5 patients in each group). Control group also consisted of 5 healthy dogs. Blood was collected on the day of admission. The diagnosis of babesiosis was confirmed by demonstration of the parasites within the infected erythrocytes in thin blood smears stained with May-Grünwald-Giemsa stain. Subspecies were confirmed using polymerase chain reaction (PCR) (Beck *et al.*, 2009).

Blood for proteomic analysis was drawn into EDTA tube (Becton Dickinson, Rutherford, NJ, USA) and centrifuged (1,300×*g* for 10 minutes) at the room temperature (RT) within 15 minutes of collection. After centrifugation, 1,100 µl EDTA plasma was separated in tubes with 0.08 mg protease inhibitors (cOmplete Ultra Tablets, Mini, EDTA-free, EASYpack, Roche Diagnostics Corp.,

Indianapolis, USA), for protection of protease degradation and centrifuged again (2,500×*g* for 15 minutes) at RT. After centrifugation plasma was immediately stored at -80 °C for further research. Aliquots of plasma from each group of study were subjected to enrichment by ProteoMiner™ beads (BioRad, Hercules, CA, USA) according to the manufacturer instructions.

Two dimensional gel electrophoresis

The immobilized pH gradient strips (pH 4-7; GE Healthcare Europe GmbH, Freiburg, Germany) were loaded with 100 µg of total protein during the rehydratation step for 20 hours according to the instructions of the manufacturer. The proteins were focused in the first dimension for a total of 68,000 Vh in the Ettan IPGphor 3 system (GE Healthcare Europe GmbH, Freiburg, Germany). After equilibration, proteins were separated in the second dimension based on their molecular weight in the Ettan DALT six (GE Healthcare Europe GmbH, Freiburg, Germany). A total of six SDS gels were run together at 5 mA per gel for 2 hours and 2 W per gel for 15 hours with the temperature of the running buffer maintained at 23 °C.

Image acquisition

Two dimensional gel were fixed and stained with Colloidal Coomassie Blue protocol (Dyballa *et al.*, 2009). The images were then captured using the ImageScanner III System (GE Healthcare Europe GmbH, Freiburg, Germany) and analyzed by specific 2D software (ImageMaster™ 2D Platinum 7.0, GE Healthcare Europe GmbH, Freiburg, Germany). Spots that were consistent in at least 3 of 5 gels in each experimental group (healthy, non-SIRS, SIRS and MODS) and showed different expression with level of significance $P<0.05$ compared with another group, were picked using Ettan spot picker from the stained gels and subjected to MS identification.

Protein identification

Proteins within gel spots were first reduced and alkylated using DTT and iodoacetamide, respectively, and then digested to peptides by trypsin proteomics grade (Sigma-Aldrich, St Louis, MO, USA) following the protocol described in Shevchenko *et al.* (2006). The tryptic peptides were analyzed by capillary reversed-phase liquid chromatography coupled online with MS/MS. The column, BioBasic-18, 5 µm particles, 300 Å pore size, 0.18 mm ID-100 mm L (Thermo, San Jose, CA), was connected to an Surveyor MS Pump Plus (Thermo, San Jose, CA) and then coupled with an ion trap mass spectrometer (LXQ, Thermo, San Jose, CA). The resulting mass spectra were searched against the *Canis familiaris* Uniprot database with the Proteome Discoverer 1.3 software (Thermo, San Jose, CA).

Results and discussion

In this study, we have investigated the proteome changes in plasma of dogs with babesiosis comparing with healthy dogs. Obtained representative 2D images of healthy dogs, dogs with SIRS, dogs with non-SIRS and dogs with MODS are shown in Figure 1.

Using proteomics approach with enrichment of low-abundance proteins before 2D electrophoresis and LC/MS/MS analysis, we have identified 22 differentially expressed proteins between healthy dogs, non-SIRS dogs with babesiosis, dogs with SIRS and dogs with MODS. Figure 2 presents the reference canine plasma map with the identified spots which were differentially expressed between groups.

Our results indicate that various physiological pathways, including acute phase response, complement and coagulation activation, lipid transport and metabolism, oxidative stress and vitamin D pathway are modulated in canine babesiosis. Some of our identified proteins could successfully discriminated patients with SIRS or MODS from healthy dogs. Further validation and confirmation for these biomarker candidates are necessary.

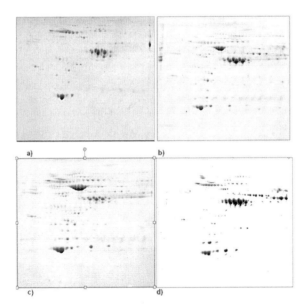

Figure 1. Representative 2-DE image of canine plasma obtained by performing the first dimension (IEF) on IPG strips pH 4-7 and the second dimension on 12.5% SDS-PAGE gels. The protein spots were visualized by colloidal coomassie staining. (a) healthy dogs; (b) dogs with non-SIRS; (c) dogs with SIRS; (d) dogs with MODS.

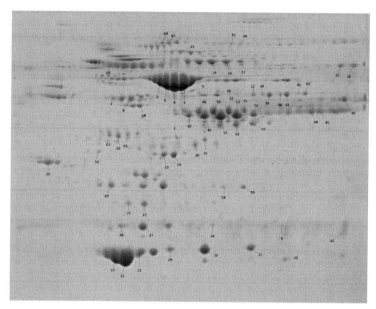

Figure 2. Reference canine plasma 2DE map with the identified numbered protein spots differentially expressed between groups.

Acknowledgements

This work was supported by a grant from the European Cooperation in Science and Technology (e-COST) through a STSM from FAP-COST 1002.

References

Beck, R., Vojta, L., Mrljak, V., Marinculić, A., Beck, A., Živičnjak, T. and Caccio, S.M., 2009. Diversity of Babesia and Theileria species in symptomatic and asymptomatic dogs in Croatia. *International Journal for Parasitology* 39: 843-848.

Dyballa, N. and Metzger, S., 2009. Fast and sensitive colloidal coomassie G-250 staining for proteins in polyacrylamide gels. Journal of Visualized Experiments 3: 1431.

Jacobson, L.S. and Clark, I.A., 1994. The pathophysiology of canine babesiosis: new approaches to an old puzzle. Journal of the South African *Veterinary Association* 65: 134-145.

Matijatko, V., Torti, M. and Schetters, T.P., 2012. Canine babesiosis in Europe: how many diseases? Trends in Parasitology 28: 99-105.

Okano, S., Yoshida, M., Fukushima, U., Higuchi, S., Takase, K. and Hagio, M., 2002. Usefulness of systemic inflammatory response syndrome criteria as an index for prognosis judgement. Veterinary Record 150: 245-246.

Shevchenko, A., Tomas, H., Havlis, J., Olsen, J.V. and Mann, M., 2006. In-gel digestion for mass spectrometric characterization of proteins and proteomes. Nature Protocols 1: 2856-60.

Taboada J. and Merchant, S. R., 1991. Babesiosis of companion animals and man. *Veterinary* Clinics of *North America*: *Small Animal Practice* 21: 103-123.

Welzl, C., Leisewitz, A.L., Jacobson, L.S., Vaughan-Scott, T. and Myburgh, E., 2001. Systemic inflammatory response syndrome and multiple-organ damage/dysfunction in complicated canine babesiosis. Journal of the South African *Veterinary Association* 72: 158-162.

Study of the effects of saturated fatty acids on the porcine epitheloid ileum IPI-2I cell line

Anna Marco-Ramell[1]*, Kerry Wallace[1], Michael Welsh[2], Gordon Allan[1], Mark Mooney[1] and Violet Beattie[3]

[1]School of Biological Sciences, Queen's University Belfast, BT9 5BN Belfast, Northern Ireland, United Kingdom; a.marco-ramell@qub.ac.uk

[2]Veterinary Sciences Division, Agri-Food & Biosciences Institut, BT4 3SD Belfast, Northern Ireland, United Kingdom

[3]Devenish Nutrition Ltd., BT3 9AR Belfast, Northern Ireland, United Kingdom

Background

A previous collaborative study between Devenish Nutrition Ltd. and QUB have shown that short and medium-chain saturated fatty acids (FAs) fed to sows during pregnancy and lactation can alter the nutritional quality and immunoglobulin profile of colostrum and milk, leading to higher growth rates of piglets and an improved immune response (unpublished data). The effect FAs in the pig still remain unclear and in order to unveil the physiological and molecular mechanisms of FA activity, several *in vitro* assays utilising the porcine epitheloid ileum IPI-2I cell line are being performed. A proteomic approach based on protein bidimensional separation coupled to mass spectrometry identification will also be carried out to profile cellular protein expression responses to these nutritional components.

In the gastrointestinal tract, L-cells in response to nutrients produce different types of hormones such as glucagon-like peptides (GLPs), derived from the specific cleavage of the proglucagon peptide. GLP-1 principally reduces food intake by modulating insulin and glucagon secretion, whilst GLP-2 stimulates intestinal growth and inhibits apoptosis of intestinal crypts (Nishimura, Hiramatsu, Monir, Takemoto, & Watanabe, 2013). Some studies have demonstrated that lipids can stimulate the production of these two hormones (Rafferty *et al.*, 2011; Sato *et al.*, 2013). In order to assess whether these GLP hormones are also produced by the IPI-2I cell line after incubation with FAs, intracellular levels of GLP-1 and GLP-2 will also be measured by immuno-based fluorescence techniques.

Material and methods

Cell culture

Ileum IPI-2I cells were cultured with DMEM-GlutaMAX™ supplemented with 10% (v/v) FCS, 50 µg/ml gentamicin and 10 µg/ml insulin and incubated in 5% CO_2-humified atmosphere at 37 °C. Cells underwent passage upon reaching 80-90% confluence and were used during described procedures between passages number 25-38.

Preparation of fatty acids

Individual saturated FAs, including butyric acid (C4), caproic acid (C6), caprylic acid (C8), capric acid (C10), lauric acid (C12), myristic acid (C14), palmitic acid (C16) and stearic acid (C18) (Sigma, St. Louis, MO) were first dissolved in ethanol and then in assay medium (culture medium containing 2% FCS) – The final ethanol concentration within was 0.01% (v/v) in all cases. Prior to addition to cell cultures all solutions were pre-warmed and control cell treatments were performed using assay medium at 0.01% ethanol.

Viability and proliferation assays

Viability and proliferation assays were performed in 96-well plates. The colorimetric thiazolyl blue tetrazolium bromide (MTT) assay, in which viable cells convert MTT (Sigma) to insoluble purple formazan, was performed for quantification of cell viability. On the other hand, the colorimetric immunoassay cell proliferation ELISA kit (Roche, Mannheim, Germany), based on the incorporation of the 5-bromo-2'-deoxyuridine (BrdU) molecule into cellular DNA, was used for cell proliferation measurement.

Proteomic analysis

Cells were seeded in flasks and incubated with FAs for 3 h and 48 h. Cells were then washed with PBS and treated with lysis buffer (7 M urea, 2 M thiourea, 4% CHAPS, 2% ampholytes, protease inhibitors). 13 cm 3-10 NL strips were rehydrated overnight with 300 µg protein in sample buffer (7 M urea, 2 M thiourea, 2% CHAPS, 1% DTT, 2% ampholytes pH 3-10, 0.002% bromophenol blue) and isoelectric separation was performed in a Multiphor II Electrophoresis Unit (GE Healthcare, Uppsala, Sweden). After focusing for a total of 17 kV/h, strips were equilibrated and separated by prepared 12.5% SDS-PAGE gels. Gels were fixed and stained with Coomassie brilliant blue. Image analysis will be performed using the Redfin 2D gel analysis image software (Ludesi, Lund, Sweden) and differential expressed spots showing a fold-change $\geq \pm 1.2$ and a $P<0.05$ will be selected for identification by mass spectrometry.

Detection of GLP gut hormones

Cells were seeded in 96 well-plates and exposed to FAs for 3 h. Cells were then fixed for 10 min with 10% formalin, permeabilized for 30 min with 0.25% Triton in PBS (PBST), blocked for 1 h with 3% BSA in PBST and incubated overnight at 4 °C with either polyclonal anti-GLP-1 or anti-GLP-2 (Santa Cruz, Dallas, TX, USA) primary antibodies diluted in 1% BSA in PBST. The following day, cells are incubated for 1 h with FITC-labelled secondary antibodies (Abcam, Cambridge, MA, USA) also diluted in 1% BSA in PBST and cell nuclei stained for 10 min with Hoechst 33258. Fluorescence was read with the ArrayScan™ XTI High Content Analysis Reader (Thermo Scientific, Waltham, MA, USA).

Results

Viability and proliferation assays

Both cell-based assay methods used to assess the *in vitro* responses to FAs exposure revealed similar effects on cellular activity after cell treatment for 48 h with short-, medium- and long-chain FAs. These assays revealed that the longer the FA chain, the more toxic the compound can be considered. Taking into account effects on cell proliferation, 0.1 mM can be considered to be the highest FA concentration suitable for further proteomic and gut hormones analyses.

Proteomic analysis

The proteomic analysis of cellular responses to exposure to FAs is on-going, with four independent experiments being performed for each experimental setup. Figure 1 illustrates a representative 2D-gel map of IPI-2I cellular proteome subjected to bidimensional electrophoresis on 13 cm 3-10NL strips and 12.5% SDS-PAGE, achieving a high resolution in the 2D separation of the cell lysate.

Detection of GLP gut hormones

Preliminary investigations have demonstrated that the treatment of IPI-2I cells with selected individual FAs increases the production of the intestinal hormone GLP-1. Further studies will examine in depth the effect of a wider range of individual Fas at different concentrations on GLP-1 and GLP-2 intracellular production.

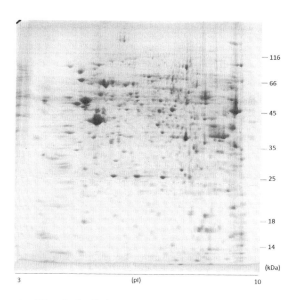

Figure 1. A representative 2D gel of cellular proteins extracted from the IPI-2I cell line.

Discussion and conclusions

Preliminary assays have shown that FAs promote cellular responses in the porcine epitheloid ileum IPI-2I cell line. On-going studies will seek to reveal new insights of the effect of FAs on intestinal metabolism using this *in vitro* model.

Acknowledgements

This work was supported by Invest Northern Ireland grant RD0513867. Authors thank Anna Gillespie (QUB) for assistance with GLP-1 analysis.

References

Nishimura, K., Hiramatsu, K., Monir, M. M., Takemoto, C., & Watanabe, T. (2013). Ultrastructural study on colocalization of glucagon-like peptide (GLP)-1 with GLP-2 in chicken intestinal L-cells. *The Journal of Veterinary Medical Science / the Japanese Society of Veterinary Science, 75*(10), 1335-9. Retrieved from http://www.pubmedcentral.nih.gov/articlerender.fcgi?artid=3942936&tool=pmcentrez&rendertype=abstract.

Rafferty, E. P., Wylie, A. R., Elliott, C. T., Chevallier, O. P., Grieve, D. J., & Green, B. D. (2011). *In vitro* and *in vivo* Effects of Natural Putative Secretagogues of Glucagon-Like Peptide-1 (GLP-1). *Scientia Pharmaceutica, 79*(3), 615-21. http://dx.doi.org/10.3797/scipharm.1104-16.

Sato, S., Hokari, R., Kurihara, C., Sato, H., Narimatsu, K., Hozumi, H., ... Miura, S. (2013). Dietary lipids and sweeteners regulate glucagon-like peptide-2 secretion. *American Journal of Physiology. Gastrointestinal and Liver Physiology, 304*(8), G708-14. http://dx.doi.org/10.1152/ajpgi.00282.2012.

Gustducin gene expression in sheep, goats and water buffalo to unravel taste signaling

Andreia T. Marques[1,2*], Ana M. Ferreira[1,3], Susana S. Araújo[1], Laura Restelli[2], Cristina Lecchi[2], André M. Almeida[1,4] and Fabrizio Ceciliani[2]

[1]*Instituto de Investigação Científica Tropical, Lisboa, Portugal and ITQB/UNL – Instituto de Tecnologia Química e Biológica, Universidade Nova de Lisboa, Portugal*

[2]*Dipartimento di Scienze Animali e Salute Pubblica, Università degli Studi di Milano, Milano, Italy; andreia.tomas@unimi.it*

[3]*Instituto de Ciências Agrárias e Ambientais Mediterrânicas (ICAAM), Universidade de Évora, Évora, Portugal*

[4]*CIISA – Centro Interdisciplinar de Investigação em Sanidade Animal, Lisboa, Portugal and IBET – Instituto de Biologia Experimental e Tecnológica, Oeiras, Portugal*

Introduction

Taste plays a crucial role in animal regulation of food intake and digestion, being a determinant factor for the choice or rejection of foods (Ginane *et al.*, 2011). In mammals, taste perception occurs via taste receptor cells (TRCs), which are clustered into taste buds and distributed across different gustatory papillae (Chandrashekar *et al.*, 2006). Bitter taste is the most interesting taste modality in herbivores as it evolved to prevent the consumption of plant toxins (Chandrashekar *et al.*, 2006). At the molecular level, bitter taste receptors (T2R) belong to the guanine nucleotide-binding regulatory protein (G protein)-coupled receptor (GPCR) super-family characterized by seven putative transmembrane domains and encoded by the TAS2R multi-gene family (Dong *et al.*, 2009).

Genetic and biochemical studies have identified specific components of the transduction pathways mediating responses to bitter compounds (Ruiz-Avila *et al.*, 1995). Alpha-gustducin is a transducin-like subunit of the G-Protein ($G_{\alpha gust}$) expressed in TRCs, which mediates bitter, sweet and *umami* gustatory signals in the taste buds of the lingual epithelium (Ruiz-Avila *et al.*, 2001). In the recent years, numerous studies on gustducin have been published but there is no structural and functional information regarding the T2R receptors coupled to gustducin in ruminants. *In vitro* and *in vivo* analysis of $G_{\alpha gust}$ knock-out mice have shown a reduced aversion to bitter compounds and reduced responses to bitter tastants in gustatory nerves (Ming *et al.*, 1998).

Data about the relation between T2R and $G_{\alpha gust}$ is still scarce in ruminants, mainly due to the lack of genomic resources. Nevertheless, cattle circumvallate taste buds have shown immunoreactivity to α-gustducin (Tabata *et al.*, 2006), whereas it was absent from non-taste cells, muscle, or brain membranes (Ming *et al.*, 1998). Outside the tongue, $G_{\alpha gust}$ expression has also been detected in TRC-like apparent chemosensory cells of the gastrointestinal tract of human, mice and rats which also express T2R receptors (Behrens and Meyerhof, 2011). To our knowledge, gastrointestinal tissues were however never tested for T2R or gustducin expression in ruminants. Considering the rather different feeding habits, and anatomy and physiology of the digestive system in these mammals, it would be interesting to analyze whether this type of signaling of bitter taste-related molecules also occurs or is a different process in non-ruminants.

Herein, we confirm the presence of Gαgust in water-buffalo papillae previously reported in cattle tissues that seem to have a high homology by qualitative and quantitative PCR. As preliminary result, we found that $G_{\alpha gust}$ does not seem to be expressed in water-buffalo and cattle's duodenum tissues.

Material and methods

Samples collection and primer design

Tissue samples (tongue containing circumvallate gustatory papillae and duodenum) were collected from four water-buffalo and cattle from a local slaughterhouse and kept at -80 °C in appropriate buffer (RNAlater). Primers were designed using Primer3 software based on the gustducin (GNAT3) mRNA sequence available in GenBank for *Bos taurus*.

RNA extraction and cDNA synthesis

Total RNA was isolated from tongue and duodenum tissues (50 mg) using Trizol standard protocol (Invitrogen), Concentration and purification of RNA were determined using spectrophotometry analysis (NanoDrop®). RNA was treated with DNase (Fermentas) and the first-strand cDNA synthesis was carried out using iScript cDNA synthesis kit (BioRad).

Qualitative mRNA expression

The cDNA from papilla was used as the template for PCRs to assert the procedure. The same primers were used in qualitative and quantitative PCR for Gustducin. PCR products were visualized on 1.6% agarose gels stained with ethidium bromide.

Results and discussion

We analyzed four papillae and four duodenums from four different water-buffalo and cattle by qualitative PCR on the cDNA of these samples. We detected gustducin expression for papillae in water-buffalo and cattle which confirm the presence of this gene in water-buffalo papillae, already reported in cattle taste tissue. An example for the expression of gustducin in papillae and duodenum is presented in Figure 1A and 1B. The conservation of T2R have already been confirm among ruminants species but this new result show us that components that interact with T2R as gustducin can also be conserved between these same species, as the primers designed for cattle gustducin work on the water-buffalo samples.

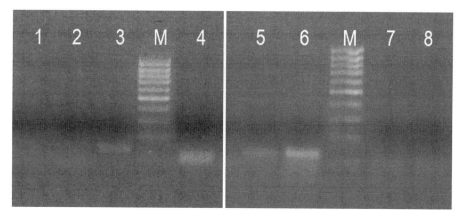

Figure 1. Example for the expression of gustducin in papillae and duodenum. Qualitative PCR result for (A) water-buffalo and (B) cattle for gustducin gene with GNAT3_RT1 primers. Lanes (M) DNA Ladder 100 bp; (1) duodenum 1; (2) duodenum 2; (3) papilla 1; (4) positive control; (5) papilla 2; (6) positive control; (7) duodenum 1; (8) duodenum 2.

No gustducin band was present either in water-buffalo and cattle's duodenum. Although gustducin expression has been detected in gastrointestinal tract of human, mice and rats, it was never tested in ruminants. We didn't know if it would be expressed in gastrointestinal tract of ruminants since they have a very different anatomy and physiology of digestive system from other non-ruminant mammals. Our results can point to different needs of bitter detection and prevention in water-buffalo, cattle than other mammal species outside the oral context. Considering the different feeding habits and digestive system of these two animals, maybe the complexity of their digestive organs can result in a no-need of extra mechanisms of avoiding plant bitter compounds consumption or of toxin removal.

Future perspectives

- RT-PCR analysis will be conducted with cattle samples.
- Western-blotting will be performed with water-buffalo and cattle samples using specific primary antibodies.
- Gustducin gene will be sequenced in water-buffalo.

References

Behrens, M., and W. Meyerhof, 2011, Gustatory and extragustatory functions of mammalian taste receptors: Physiology & Behavior, v. 105, p. 4-13.

Chandrashekar, J., M. A. Hoon, N. J. P. Ryba, and C. S. Zuker, 2006, The receptors and cells for mammalian taste: Nature, v. 444, p. 288-294.

Dong, D., G. Jones, and S. Zhang, 2009, Dynamic evolution of bitter taste receptor genes in vertebrates: Bmc Evolutionary Biology, v. 9.

Ginane, C., R. Baumont, and A. Favreau-Peigne, 2011, Perception and hedonic value of basic tastes in domestic ruminants: Physiology & Behavior, v. 104, p. 666-674.

Ming, D., L. Ruiz-Avila, and R. F. Margolskee, 1998, Characterization and solubilization of bitter-responsive receptors that couple to gustducin: Proceedings of the National Academy of Sciences of the United States of America, v. 95, p. 8933-8938.

Ruiz-Avila, L., G. T. Wong, S. Damak, and R. F. Margolskee, 2001, Dominant loss of responsiveness to sweet and bitter compounds caused by a single mutation in alpha-gustducin: Proceedings of the National Academy of Sciences of the United States of America, v. 98, p. 8868-8873.

Ruiz-Avila, L., S. K. McLaughlin, D. Wildman, P. J. McKinnon, A. Robichon, N. Spickofsky, and R. F. Margolskee, 1995, Coupling of bitter receptor to phosphodiesterase though transducin taste receptor cells: Nature, v. 376, p. 80-85.

Tabata, S., K.-i. Kudo, A. Wada-Takemura, S. Nishimura, and H. Iwamoto, 2006, Structure of bovine fungiform taste buds and their immunoreactivity for gustducin: Journal of Veterinary Medical Science, v. 68, p. 953-957.

Molecular studies on adipose tissue in turkey: searching for welfare biomarkers

Andreia T. Marques[1]*, Sara Rigamonti[1], Cristina Lecchi[1], Guido Grilli[1], Sara Rota Nodari[2], Leonardo James Vinco[2] and Fabrizio Ceciliani[1]
[1]Dipartimento di Scienze Animali e Salute Pubblica, Università degli Studi di Milano, Milano, Italy; andreia.tomas@unimi.it
[2]National Reference Centre of Animal Welfare, Istituto Zooprofilattico Sperimentale della Lombardia e dell'Emilia Romagna B. Ubertini, Brescia, Italy

Introduction

Animal welfare is a major consideration in animal productions, in order to produce safe and quality food. Road transportation of animals is an inevitable practice that animals encounter in the livestock industry and represents a critical phase in animal meat production (Minka, 2006; Schwartzkopf-Genswein *et al.*, 2012).

Recently, it has been suggested that acute phase proteins (APPs) may also be useful in the assessment of animal welfare, such as stress (Murata *et al.*, 2004), including road transport (Giannetto *et al.*, 2011), indicating their potential use as an objective parameter to evaluate stress.

Although APPs are produced mainly in liver, it has been recently shown that they can be produced also in adipose tissue (Deng and Scherer, 2010).

Adipose tissue is a loose connective tissue which is involved in homeostasis regulation and it is increasingly recognized as an active endocrine organ regulating metabolic, reproduction, oxidative stress, inflammatory pathways, as well as innate and adaptive immune response. Moreover, during a stress condition, the activation of the sympatho-adrenal and hypothalamic-pituitary-adrenal axis leads in turn to the activation of adipose tissue metabolism (Charmandari *et al.*, 2005) and, in humans, to the activation and release of APPs (Murata, 2007). Therefore, adipose tissue has been proposed to be extremely reactive and sensitive to stress.

While APPs have been thoughtfully studied in mammals, no information about APPs and its relationship with adipose tissue in poultry has been made available so far. The aim of this research project is to find markers during the transport related – Acute Phase Response (APR) in turkey (*Meleagris gallopavo*).

Herein, we identified three major APPs on turkey – α1-acid glycoprotein (AGP), serum amyloid A (SAA) and PIT54 produced in liver and adipose tissue by mRNA expression analyses.

Material and methods

Samples collection

Turkey liver and adipose tissue were collected from five healthy not-transported animals and five transported animals. Portions of liver and adipose tissue were removed immediately after slaughtering and collected into liquid nitrogen and stored at -80 °C.

Primers design and RNA extraction and cDNA synthesis

Three APPs were selected (SAA, AGP and PIT54) based on previous studies on chicken. Three housekeeping genes were selected (GAPDH, RPL4 and YWHAZ) based on previous studies and literature. Total RNA was isolated from liver and adipose tissue using Trizol standard protocol (Invitrogen). Concentration and purification of RNA were determined using spectrophotometry analysis (NanoDrop®). RNA was treated with DNase (Fermentas) and the first-strand cDNA synthesis was carried out using iScript cDNA synthesis kit (BioRad).

Qualitative and quantitative mRNA expression

The same primers were used in qualitative and quantitative PCR for SAA, AGP and PIT54. Each sample was tested in duplicate. The thermal profile was 50 °C for 2 min, 95 °C for 10 min, 40 cycles of 95 °C for 8 s and 60 °C for 20 s. Conditions for melting curve construction were 95 °C for 5 s, decreasing to 55 °C for 5 s and increasing to 95 °C for 5 s. Results were compared using the Δ-Δ Cq method.

Western blot analysis

Samples for Western blot analysis were prepared from aliquots of 50-100 mg tissues. Aliquots with different concentrations were separated by 12% sodium dodecyl sulphate–polyacrylamide gel electrophoresis (SDS–PAGE) and Western blotted onto nitrocellulose membrane. The membranes were immunolabelled for the presence of AGP using bovine AGP antibody and immunoreactive bands were visualized by enhanced chemiluminescence (ECL) using Immobilon Western Chemiluminescence substrate (Millipore). Purified AGP from bovine was used as positive control.

Results and discussion

In this work, we have identified three major APPs on turkey – α1-acid glycoprotein (AGP), serum amyloid A (SAA) and PIT54 produced in liver by mRNA expression analyses (Figure 1). Adipose tissue can also produce a small quantity of these APPs in a non-stressed animal, which surprisingly increases after exposure to a stress condition. Quantitative PCR investigation revealed that AGP, PIT54 and SAA seem to be down regulated on liver after an exposure to a stress condition, such

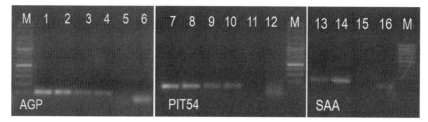

Figure 1. Qualitative PCR analysis for AGP, PIT54 and SAA genes. Qualitative PCR result for AGP, PIT54 and SAA genes with liver and adipose tissue samples from transported animals. Lanes (M) DNA Ladder 100 bp; (1) liver 1; (2) liver 2; (3) adipose tissue 1; (4) adipose tissue 2; (5) negative control; (6) positive control; (7) liver 1; (8) liver 2; (9) adipose tissue 1; (10) adipose tissue 2; (11) negative control; (12) positive control; (13) liver 1; (14) adipose tissue 1; (15) negative control; (16) positive control.

as road transportation (Figure 2A, 2C and 2E). Interestingly, these same APPs had their expression highly increased on adipose tissue after road transportation (Figure 2B, 2D and 2F).

The second part of the experiment was focused on proteins. We demonstrated that bovine anti-AGP cross-reacted with the equivalent protein of turkey (Figure 3). As preliminary result, we detected many AGP isoforms produced by liver but in adipose tissue only a high molecular weight of AGP seems to be present. Immunohistochemistry studies will be done to detect the precise AGP location in liver and adipose tissue, as well as proteomic studies, in order to identify possible differences in glycosylation pattern of AGP, or to extend the investigation to other possible candidates.

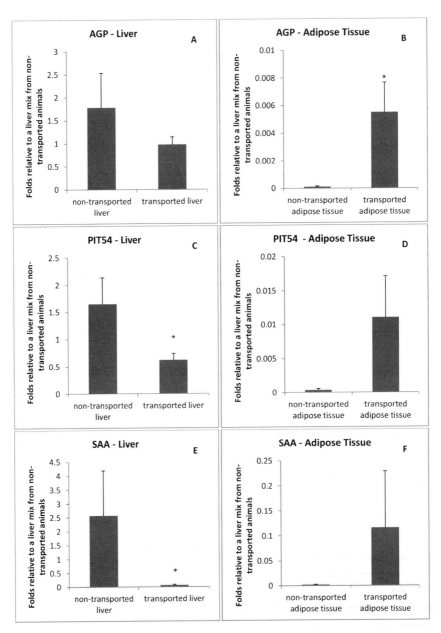

Figure 2. Quantitative PCR analysis for AGP, PIT54 and SAA. Relative AGP, PIT54 and SAA gene expression in liver and adipose tissue from transported and no-transported turkey samples by qPCR. The results were normalized using GAPDH, RPL4 and YWHAZ. A liver mix from all no-transported samples was used as reference sample and data are means of 5 different animals. * differences are statistically significant (P<0.05).

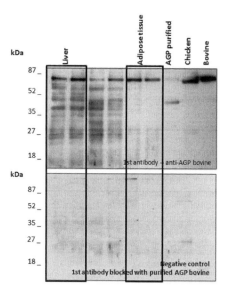

Figure 3. Testing cross-reactivity of anti-bovine AGP against turkey AGP. SDS–PAGE and Western blot analysis of turkey liver and adipose tissue samples. AGP was identified on turkey samples using an anti-bovine AGP antibody. Purified AGP from bovine was used as positive control.

References

Charmandari, E., C. Tsigos, and G. Chrousos, 2005, Endocrinology of the stress response, Annual Review of Physiology: Annual Review of Physiology, v. 67: Palo Alto, Annual Reviews, p. 259-284.

Deng, Y. F., and P. E. Scherer, 2010, Adipokines as novel biomarkers and regulators of the metabolic syndrome, *in* A. C. Powers, and R. S. Ahima, eds., Year in Diabetes and Obesity: Annals of the New York Academy of Sciences, v. 1212: Malden, Wiley-Blackwell, p. E1-E19.

Eckersall, P. D., 2000, Recent advances and future prospects for the use of acute phase proteins as markers of disease in animals: Revue De Medecine Veterinaire, v. 151, p. 577-584.

Giannetto, C., F. Fazio, S. Casella, S. Marafioti, E. Giudice, and G. Piccione, 2011, Acute phase protein response during road transportation and lairage at a slaughterhouse in feedlot beef cattle: J Vet Med Sci, v. 73, p. 1531-4.

Minka, N. S., Ayo J.O., 2006, Effects of loading behaviour and road transport stress on traumatic injuries in cattle transported by road during the hot-dry season: Livestock Science, v. 107, p. 91-95.

Murata, H., 2007, Stress and acute phase protein response: an inconspicuous but essential linkage: Vet J, v. 173, p. 473-4.

Murata, H., N. Shimada, and M. Yoshioka, 2004, Current research on acute phase proteins in veterinary diagnosis: an overview: Veterinary Journal, v. 168, p. 28-40.

Salamano, G., E. Mellia, M. Tarantola, M. S. Gennero, L. Doglione, and A. Schiavone, 2010, Acute phase proteins and heterophi:lymphocyte ratio in laying hens in different housing systems: Veterinary Record, v. 167, p. 749-751.

Schwartzkopf-Genswein, K. S., L. Faucitano, S. Dadgar, P. Shand, L. A. González, and T. G. Crowe, 2012, Road transport of cattle, swine and poultry in North America and its impact on animal welfare, carcass and meat quality: a review: Meat Sci, v. 92, p. 227-43.

Modelling of 3-oxoacyl syntheses 2 from *Brucella suis* to be used for structure based drug design

Jani Mavromati[1]*, Dimitrios Vlachakis[2], Sophia Kossida[2] and Xhelil Koleci[1]
[1]*Veterinary Public Health Department, Veterinary Medicine Faculty, Agricultural University of Tirana, Albania; j.mavromati@hotmail.com*
[2]*Biomedical Research Foundation of the Academy of Athens, Greece*

Introduction

The β-ketoacyl carrier protein synthases (β-KAS) are a key regulator of fatty acid biosynthesis, which has emerged as a target for the development of therapeutic agents (Komaki-Yasude *et al.*, 2013). Thiolactomycin and Cerulenin are two natural ligands, capable of binding to the active site of β-KAS enzymes and inhibit their function (Komaki-Yasude *et al.*, 2013). A variety of different structures of these enzymes, from different origins, have already been determined from X-Ray and NMR studies. There are two main fatty acid synthesis systems (Komaki-Yasude *et al.*, 2013). The type I system, is found in the most complex organisms, such as mammals. This system composes a unique and multifunctional polypeptide. The other, the type II or dissociated system, is found in lower organisms such as bacteria and plants. Three different synthases have been found to regulate the pathways of the type II system (KAS I, II and III). According to work done by Zhang *et al.* (2000), the KAS enzymes are mainly known for catalysing the Claisen condensation reaction. This is done by the transport of an acyl primer to malonyl-ACP. The result is therefore a β-Ketoacyl-Acyl carrier protein elongated by two carbon units.

Coordinate preparation and sequence alignment

3D coordinates were obtained from the X-ray solved, crystal structures with RCSB codes: 3LRF.

The amino acid sequence of the 3-oxoacyl-[acyl-carrier-protein]synthase 2 was obtained from the GenBank database (accession no: WP_006195353.1, entry name: 3-oxoacyl-[acyl-carrier-protein] synthase 2 [*Brucella suis*]). Using the Gapped-BLAST (Altschul *et al.*, 2004) through NCBI (Benson *et al.*, 2007) the homologous species of the *Brucella melitensis* helicase was identified, which was used as template for the homology modelling. The sequence alignment was done using the online version of ClustalW (Thomson *et al.*, 1994). The alignment was repeated using Hidden Markov Models and the result was the same as the one obtained by ClustalW, due to the fact that there are several anchoring conserved motifs throughout the alignment (Eddy, 1995).

Identities 176/431 = 41%, Positives 249/431 = 57%, Gaps 39/431 = 9%

Query	22	MRRVVVTGMGIVSSIGSNTEEVTASLREAKSGISRAEEYAELGFRCQV----------HG	71
		MRRVV+TG+G+VS + S EE L +SG R E+ CQ+ +G	
Sbjct	1	MRRVVITGLGLVSPLASGVEETWKRLLAGESGARRVTEFEVDDLACQIACRIPVGDGTNG	60
Query	72	A--PDIDIESLVDRRAMRFHGRGTAWNHIAMDQAIADAGLTEEEVSNE-RTGIIMGSGGP	128
		PD+ ++ R+ F + A DQA+ DAG E ++ RTG+++GSG	
Sbjct	61	TFNPDLHMDPKEQRKVDPF----IVYAVGAADQALDDAGWHPENDEDQVRTGVLIGSGIG	116
Query	129	STRTIVDSADITREKGPKRVGPFAVPKAMSSTASATLATFFKIKGINYSISSACATSNHC	188
		IV++ R+KGP+R+ PF +P + + AS ++ K++G N+S+ +ACAT H	
Sbjct	117	GIEGIVEAGYTLRDKGPRRISPFFIPGRLINLASGHVSIKHKLRGPNHSVVTACATGTHA	176
Query	189	IGNAYEMIQYGKQDRMFAGGCEDLDWTLSVL-FDAMGAMSSKYNDTPSTASRAYDKNRDG	247
		IG+A +I +G D M AGG E +S+ F A A+S+K ND P+ ASR YD +RDG	
Sbjct	177	IGDAARLIAFGDADVMVAGGTESPVSRISLAGFAACKALSTKRNDDPTAASRPYDGDRDG	236
Query	248	FVIAGGAGVLVLEDLETALARGAKIYGEIVGYGATSDGYDMVAP--SGEGAIRCMKMALS	305
		FV+ GAG++VLE+LE ALARGAKIY E++GYG + D + + AP SGEGA RCM AL	
Sbjct	237	FVMGEGAGIVVLEELEHALARGAKIYAEVIGYGYGMSGDAFHITAPTESGEGAQRCMVAALK	296
Query	306	TVTSKIDYINPHATSTPAGDAPEIEAIRQIFGAGDVCPPIAATKSLTGHSLGATGVQ	362
		V +IDYIN H TST A D E+ A+ ++ G +++TKS GH LGA G	
Sbjct	297	RAGIVPDEIDYINAHGTSTMA-DTIELGAVERVVGEAAAKISMSSTKSSIGHLLGAAGAA	355
Query	363	EAIYSLLMMQNNFICESAHIEELDPAFADMPIVRKRIDNV-------QLNTVLSNSFGFG	415
		EAI+S L +++N + ++ D P + RID V +++ LSNSFGFG	
Sbjct	356	EAIFSTLAIRDNIAPATLNL--------DNPAAQTRIDLVPHKPCERKIDVALSNSFGFG	407
Query	416	GTNATLVFQRY	426
		GTNA+LV +RY	
Sbjct	408	GTNASLVLRRY	418

Homology modelling

The homology modelling of the 3-oxoacyl-[acyl-carrier-protein] synthase was carried out using the Modeller package (version 9.10) (Sali *et al.*, 1995). The RCSB entry 3LRF was used as template structure. The sequence alignment between the raw sequence of the 3-oxoacyl-[acyl-carrier-protein] synthase 2 and the full sequence of the beta-ketoacyl synthase from *B. melitensis* revealed almost 40% identity. The abovementioned identity score is on the lower boundary at which conventional homology modelling techniques are applicable. The homology model method of Modeller comprises the following steps: First, an initial partial geometry specification, where an initial partial geometry for each target sequence is copied from regions of one or more template chains. Secondly, the insertions and deletions task, where residues that still have no assigned backbone coordinates are modelled. Third step is the loop selection and sidechain packing, where a collection of independent models is created. Finally, necessary secondary structure predictions were performed using the NPS (Network Protein Sequence Analysis) web-server and the GeneSilicoMetaServer, which confirmed the choice of the beta-ketoacyl synthase from *B. melitensis* as template for this study (http://npsa-pbil.ibcp.fr).

Molecular electrostatic potential

Electrostatic potential surfaces were calculated by solving the nonlinear Poisson-Boltzmann equation using finite difference method as implemented in the PyMOL Software (Delano, 2002).

Energy minimization and molecular dynamics simulations

Energy minimizations were used to remove any residual geometrical strain in each molecular system of the 3-oxoacyl-[acyl-carrier-protein] synthase 2 from *B. suis*, using the Charmm27 forcefield as it is implemented into the Gromacs suite, version 4.5.5 (Hess *et al.*, 2008).

Model evaluation

Evaluation of the model quality and reliability in terms of its 3D structural conformation is very crucial for the viability of this study. Therefore, the produced model was initially evaluated within the Gromacs package by a residue packing quality function, which depends on the number of buried non-polar side chain groups and on hydrogen bonding. Moreover, the suite PROCHECK (Laskowski *et al.*, 1996) was employed to further evaluate the quality of the produced 3-oxoacyl-[acyl-carrier-protein] synthase 2 from *B. suis* model. Verify3D (Eisenberg *et al.*, 1997) was also used to evaluate whether the model of the 3-oxoacyl-[acyl-carrier-protein] synthase 2 from *B. suis* is similar to known protein structures. Finally, the Molecular Operating Environment (MOE) suite was used to evaluate the 3D geometry of the models in terms of their Ramachandran plots, omega torsion profiles, phi/psi angles, planarity, C-beta torsion angles and rotamer strain energy profiles.

Acknowledgements

I would like to thank COST FA1002 for making this STSM possible and Biomedical research foundation of the academy Athens, Greece for hosting me for this STSM.

References

Benson, D. A., Karsch-Mizrachi, I., Lipman, D. J., Ostell, J. and Wheeler, D. L. (2007). GenBank. Nucleic Acids Res. 35, 21-25.

DeLano WL (2002) ThePyMOL User's Manual San Carlos, CA, USA.: DeLano Scientific.

Eddy SR. Multiple alignment using hidden Markov models. ProcIntConfIntellSystMol Biol. 1995;3:114-20

Eisenberg, D., R. Luthy, and J.U. Bowie, VERIFY3D: assessment of protein models with three-dimensional profiles. Methods Enzymol 277 (1997) 396-404

Hess, B., C. Kutzner, D. van der Spoel and E. Lindahl, GROMACS 4: algorithms for highly efficient, load-balanced, and scalable molecular simulation. J. Chem. Theory Comput. 4 (2008) 435-447.

Komaki-Yasuda K, Okuwaki M, Nagata K, Kawazu S, Kano S. Identification of a Novel and Unique Transcription Factor in the Intraerythrocytic Stage of Plasmodium falciparum. PLoS One. 2013 Sep 5;8(9):e74701.

Laskowski, R.A., J.A. Rullmannn, M.W. MacArthur, R. Kaptein, and J.M. Thornton, AQUA and PROCHECK-NMR: programs for checking the quality of protein structures solved by NMR. J Biomol NMR 8 (1996) 477-86.

Sali A, Potterton L, Yuan F, van Vlijmen H, Karplus M (1995) Evaluation of comparative protein modeling by MODELLER. Proteins 23: 318-326.

Thompson, J. D., Higgins, D. G. and Gibson, T. J. (1994). CLUSTAL W: improving the sensitivity of progressive multiple sequence alignment through sequence weighting, position specific gap penalties and weight matrix choice. Nucleic Acids Res. 22, 4673-4680.

Zhang YM, Rao MS, Heath RJ, Price AC, Olson AJ, Rock CO, White SW. Identification and analysis of the acyl carrier protein (ACP) docking site on beta-ketoacyl-ACP synthase III. J Biol Chem. 2001 Mar 16;276(11):8231-8.

Proteomic analysis of gilthead sea bream plasma with amyloodiniosis

Márcio Moreira[1]*, Denise Schrama[2], Florbela Soares[1], Pedro Pousão-Ferreira[1] and Pedro Rodrigues[2]
[1]Aquaculture Research Centre, National Institute for the Sea and Atmosphere (IPMA), Av. 5 de Outubro s/n, 8700-305 Olhão, Portugal; marciomoreira.27@gmail.com
[2]CCMAR – Centre of Marine Sciences, University of Algarve, Campus de Gambelas, 8005-139 Faro, Portugal

Introduction

Diseases are one of the most important constraints in aquaculture, especially in intensive fish farming, limiting their success. The open design of many of the aquaculture systems also allows the transmission of infectious pathogens where they find ideal condition to cause a disease outbreak (Balcázar, 2006; Mladineo, 2006).

In Southern Europe, amyloodiniosis represents a major bottleneck for semi-intensive aquaculture, and is one of the most serious impediments to warm water aquaculture, with over 100 species known to be susceptible (Noga, 1996).

This disease is caused by one of the most common and important parasitic dinoflagellate in fish, *Amyloodinium ocellatum*. This parasite can affect almost all fish that live within its ecological range, causing serious morbidity and mortality in brackish and marine warm water fish in different aquaculture facilities worldwide, and is often considered the most consequential pathogen of marine fish (Paperna *et al.*, 1981). It is a 'quiet' disease since the outbreaks occur extremely rapid, and usually by the time of its detection, contaminated fish no longer respond to treatment, resulting in 100% mortality in a few days (Soares *et al.*, 2011).

The symptomatology of this disease is characterized by changes in fish behaviour, with jerky movements, swimming at the water surface and decreased appetite (Soares *et al.*, 2011). These may include increased respiratory rate and gathering at the surface or in areas with higher dissolved oxygen concentrations. Information regarding the physiological responses from the host to the parasite infestation are scarce.

Proteomics is one of the new approaches to understand fish diseases, which can elucidate the functional responses of organisms to environmental and biological challenges, thus facilitating the identification of disease biomarkers (Alves *et al.*, 2010; Cox and Mann, 2011).

In this work we will analyse the proteomic profile of gilthead sea bream (*Sparus aurata*, Linnaeus, 1758) plasma, one of the most important cultured fishes in Southern Europe (Soares *et al.*, 2011).

Plasma proteome is a unique biomarker source since it reflects the global expression of all cellular genomes (Isani *et al.*, 2011).

Material and methods

Ninety gilthead sea breams, with a mean body weight of 87.2 ± 17 g, were transferred to 200 L rectangular plastic tanks (Control, T1 – with *A. ocellatum*, T2 – with *A. ocellatum* and other ectoparasites) in duplicate, at EPPO – IPMA Aquaculture Research Station, (Olhão, Portugal). The fish were kept at 22±0.2 °C, in closed recirculation seawater systems, artificial aeration and 24 h light photoperiod. When the contamination reached 500 parasites per branchial arc, 4 fish from each tank were anesthetized with 2-phenoxyethanol (Sigma Aldrich), and approximately 1 ml of blood was withdrawn using syringes heparinized with 1% EDTA. The blood was centrifuged at 2500 rpm for 10 minutes and plasma was collected and kept at -80 °C for subsequent analysis. After that fish were slaughtered with an overdose of anesthetic, measured and weighted. Plasma proteins were quantified by the Bradford method and 50 µg were labeled with either CyDye3 or CyDye5. A pool of samples was labeled with CyDye2 as internal standard. After an incubating period of half hour, the reaction was stopped by adding Lysine (10 mM). Plasma proteins were focused on 24 cm Immobiline DryStrips (GE Healthcare, Sweden) with pH 4-7 until a total of 60,346 Vhr. A reduction and alkylation step was done with DTT and iodoacetamide, respectively before separation according to molecular weight on 12.5% polyacrylamide gels.

Gels were scanned on a 9400 Typhoon scanner (GE Healthcare, Sweden) and analyzed with SameSpots (Totallab, United Kingdom) with filters for average normalized volume ≤10,000 and spot area ≤500, Significantly different spots ($P<0.05$ by ANOVA and a frequency discovery rate (FDR)) were excised manually and sequenced by LC/MS MS.

Results and discussion

After gel analysis 409 spots were differently expressed between treatments (over 2078 spots detected). The gel is shown in Figure 1.

Of the 409 spots we selected for protein identification the spots that presented differences between T1 and the other treatments, which could possibly be specific markers of the physiological responses of the host to *A. ocellatum*. We have also selected 6 spots that had volume differences between T1 and T2 versus control that could represent some physiological responses to branchial parasites. The results obtained from the sequencing are in the Table 1.

A decrease in the expression of Apolipoprotein A-I was observed in T1. This protein is produced in the liver and is involved in lipid absorption in the intestine (Concha *et al.*, 2005) and with a possible role in innate immunity (Concha *et al.*, 2003; Concha *et al.*, 2004). This variation in its expression might indicate an inhibition effect of the parasite *A. ocellatum*.

The overexpression of Transferrin, a protein with a central role in iron metabolism, and associated with a role in the innate immune system response (García-Fernández *et al.*, 2011), can be explained as an immune response against the parasite *A. ocellatum* and be linked to a higher requirement of oxygen by the host.

The uncharacterized protein OS Takifugu rubripes GN LOC101068102 PE 4 SV 1, in which the closest ID by BLAST (http://www.uniprot.org/blast/uniprot/2014090493DNYK5ZU1) is fibrinogen beta chain protein, plays an important role with fibrin in blood clotting, fibrinolysis, cellular and matrix interactions, inflammation, wound healing, neoplasia and can have an immune role in the liver of fishes (Xie *et al.*, 2009). The overexpression of this protein in T1 and T2 can be related to a response to the lesions produced by the ectoparasites in the branchia of gilthead sea bream.

The tr|Q4RIP0|Q4RIP0_TETNG, tr|D0VB93|D0VB93_SPAAU and tr|W5N831|W5N831_LEPOC are peptides with unknown function.

Overall, this preliminary study revealed that amyloodiniosis can upregulate or downregulate proteins with known functions on the immune response, and physiological responses related with inflammation and iron transport of the host to the parasite. Further interdisciplinary studies will be needed to elucidate the role of these proteins in an amyloodiniosis outbreak, like stress reactions and possible histopathological changes in the host organs.

References

Alves, R.N., Cordeiro, O., Silva, T.S., Richard, N., de Vareilles, M., Marino, G., Di Marco, P., Rodrigues, P.M., and Conceição, L.E.C., 2010. Metabolic molecular indicators of chronic stress in gilthead seabream (*Sparus aurata*) using comparative proteomics. Aquaculture 299, 57-66;

Balcázar, J. L., Blas, I. D., I. Ruiz-Zarzuela, I., Cunningham, D., Vendrell, D. and Múzquiz, J. L., 2006. The role of probiotics in aquaculture. Veterinary Microbiology 114 (3-4), 173-186;

Concha, M.I., López, R., Villanueva, J., Báez, N., and Amthauer, R., 2005. Undetectable apolipoprotein A-I gene expression suggests an unusual mechanism of dietary lipid mobilisation in the intestine of *Cyprinus carpio*. The Journal of Experimental Biology 208, 1393-1399;

Concha, M.I., Molina, S., Oyarzún, C., Villanueva, J., and Amthauer, R., 2003. Local expression of apolipoprotein A-I gene and a possible role for HDL in primary defence in the carp skin. Fish & Shellfish Immunology 14, 259-273;

Concha, M.I., Smith, V.J., Castro, K., Bastías, A., Romero, A., and Amthauer, R., 2004. Apolipoproteins A-I and A-II are potentially important effectors of innate immunity in the teleost fish *Cyprinus carpio*. European Journal of Biochemistry 271, 2984-2990;

Cox, J., and Mann, M., 2011. Quantitative, High-Resolution Proteomics for Data-Driven Systems Biology. Annual Review of Biochemistry 80, 273-299;

Esteban, M.Á., 2012. An Overview of the Immunological Defenses in Fish Skin. International Scholarly Research Network Immunology 2012, 29 pp.;

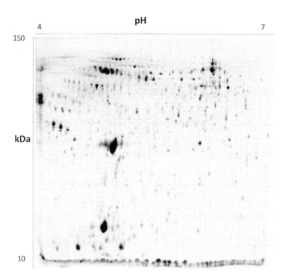

*Figure 1. DIGE gel obtained from gilthead sea bream (*Sparus aurata*) plasma.*

Table 1. Results of the sequencing and expression patterns of the selected spots.

Accession	Description	Expression
tr|O42175|APOA1_SPAAU	Apolipoprotein A I OS Sparus aurata GN apoa1 PE 2 SV 1	T1<T2, Control
tr|Q4RIP0|Q4RIP0_TETNG	Chromosome 7 SCAF15042 whole genome shotgun sequence Fragment OS Tetraodon nigroviridis GN GSTENG0003380800	T1>T2, Control
tr|D0VB93|D0VB93_SPAAU	Estrogen-regulated protein OS Sparus aurata PE 2 SV 1	T1, T2>Control
tr|F2YLA1|F2YLA1_SPAAU	Transferrin OS Sparus aurata GN Tf PE 2 SV 1	T1>T2, Control
tr|W5N831|W5N831_LEPOC	Uncharacterized protein OS Lepisosteus oculatus GN KRT9 2 of 3 PE 3 SV 1	T1, T2<Control
tr|H2S183|H2S183_TAKRU	Uncharacterized protein OS Takifugu rubripes GN LOC101068102 PE 4 SV 1	T1, T2>Control

García-Fernández, C., Sánchez, J.A., and Blanco, G., 2011. Characterization of the gilthead seabream (*Sparus aurata* L.) transferrin gene: Genomic structure, constitutive expression and SNP variation. Fish & Shellfish Immunology 31, 548-556;

Isani, G., Andreani, G., Carpenè, E., Di Molfetta, S., Eletto, D., and Spisni, E., 2011. Effects of waterborne Cu exposure in gilthead sea bream (*Sparus aurata*): A proteomic approach. Fish & Shellfish Immunology 31, 1051-1058;

Mladineo, I., 2006. Parasites of Adriatic cage reared fish. Acta Adriatica 47, 23-28;

Noga EJ., 1996. Fish Disease: Diagnosis and Treatment. Iowa State University Press, Ames, Iowa, 376 pp.;

Paperna, I., Colorni, A., Ross, B. and Colorni, B., 1981. Diseases of marine fish cultured in Eilat mariculture project based at the Gulf of Aqaba, Red Sea. European Mariculture Society Special Publication 6, 81-91;

Soares, F., Quental Ferreira, H., Cunha, E., and Pousão-Ferreira, P., 2011. Occurrence of *Amyloodinium ocellatum* in aquaculture fish production: a serious problem in semi-intensive earthen ponds, Aquaculture Europe 36 (4), 13-16;

Xie, F.J., Zhang, Z.P., Lin, P., Wang, S.H., Zou, Z.H., and Wang, Y.L., 2009. Identification of immune responsible fibrinogen beta chain in the liver of large yellow croaker using a modified annealing control primer system. Fish & Shellfish Immunology 27, 202-209;

The measurement of chicken acute phase proteins using a quantitative proteomic approach

Emily L. O'Reilly[1]*, P. David Eckersall[1], Gabriel Mazzucchelli[2] and Edwin De Pauw[2]

[1]Institute of Biodiversity, Animal Health and Comparative Medicine, College of Medicine, Veterinary and Life Sciences, Glasgow University, Glasgow, Scotland, United Kingdom; emilymoo90@hotmail.com
[2]Mass Spectrometry Laboratory, GIGA-Research, Department of Chemistry, University of Liège, Liège, Belgium

Introduction

The purpose of this study was to establish a method, using selected reaction monitoring (SRM), to measure known acute phase proteins (APPs) serum amyloid A (SAA), C-reactive protein (CRP), ovotransferrin (OVT), transthyretin (TTN) and apolipoprotein A-1 (Apo-A1) in chickens. The characterisation and measurement of APPs in chickens is a challenging area owing to the limited availability of specific antisera and in the case of SAA and CRP, the low abundance in serum. Previous methods of measuring these APPs have used immunoassays such as ELISAs or biochemical methods. This investigation aimed to take a proteomic approach to identify and quantify these APPs in chicken serum. To develop a targeted SRM method to measure multiple proteins within a single sample, the APPs were firstly identified in a initial shotgun (SG) analysis. In the case of the low abundant proteins SAA and CRP, enrichment steps were undertaken prior to SG analysis to enrich these proteins within a sample. The proteins within the serum samples were separated by molecular weight, to increase the relative concentrations of SAA and CRP (low molecular weight proteins) within the serum. Affinity chromatography using phosphoyl choline, to which CRP is known to bind with high affinity was also undertaken with the aim of enriching the protein. The other APPs were identified successfully in SG analysis of the whole serum and no enrichment was needed. Initial SG analysis and measurement of the APPs was undertaken on three groups of birds: highly acute phase viraemic birds, acute phase birds and a non-acute phase group. For the SG analysis the samples for the three groups were pooled. The SRM targeted approach was undertaken on individual samples from each group.

Materials and methods

Highly acute phase serum from viraemic birds (n=3), acute phase serum (n=4) and non-acute phase serum (n=4), were designated as groups 1-3 and samples from each group were pooled together for the SG analysis. An enrichment step, to fractionate the lower molecular weight proteins (<30 kDa) was undertaken using Protein Discovery GelFree® 8100 Fractional system, with a 10% cartridge. For CRP, highly acute phase and acute phase serum were added to four mini-columns containing immobilised p-aminophenyl phosphoyl choline gel (Thermo Scientific 20307). Columns were washed and serum added twice further before a final wash stage and elution. The eluates were pooled and concentrated. The low molecular weight fractions (<30 kDa) and the phosphoyl column eluates, together with whole serum pooled into groups 1-3 underwent sample preparation that included trypsin digestion, as described in Collodoro, *et al.*, (2012). The peptides were applied over 2D nano UPLC system (RP × RP, nano Acquity, Waters) coupled to a Q-Exactive Hybrid Quadrupole–Orbitrap mass spectrometer (Thermo Scientific) with a top 12 data dependant acquisition method (Thermo Scientific). The Q-Exactive-Orbitrap Mass spectrometer (Thermo) then performed MS/MS spectra analysis, and comparisons were made to the *Gallus gallus* protein database (Uniprot) using Sequest (Thermo Scientific) and Mascot (Matrix Science). A quantitative targeted method was established by identifying acute phase proteins within the SG results and characterising suitably quantotypic peptides from each target protein. Once peptides were identified and validated the nano UPLC – Q-Exactive was used in the targeted mode (single ion monitoring SIM/targeted MS2). Data was assimilated and analysed using Skyline software and the peptide intensities (the sum of all the fragment ions of that peptide) were used to determine the intensities of each APP in each sample. For each APP, three peptide intensities were used and one way ANOVA and t-tests (Microsoft Excel, 2010) were used to compare the peptide intensity for each selected peptide across the three sample groups and identify significant differences.

Results

C-reactive protein was not identified during SG analysis of the whole serum, the lower molecular weight fractions or the phosphyl choline elutions and as such was not quantified or investigated further. Enriching the lower molecular weight proteins within the serum by fractionating according to molecular weight resulted in the identification of a higher number of lower molecular weight proteins and increased the number of SAA peptides identified. Target peptides were selected for the APPs of interest and the targeted method determined the peptide intensities for each APP for each of the 11 samples. Table 1 details the target peptides for each APP. The peptide intensities for each group were statistically evaluated and the p-values for each peptide are also detailed in Table 1.

For SAA the highly acute phase group was the only group to identify all three of the SAA peptides. No SAA peptides were identified in the non-acute phase group and a single peptide was identified in one sample of the acute phase group only. Similarly OVT also appears to behave as a positive APP, though it was detectable in all of the samples analysed (Figure 1). Ovotransferrin had significantly higher intensities in highly acute phase group compared to the acute phase and control groups in two of the three peptides only. For TTN one of the three peptides was significantly lower in the highly acute phase but also low in the other groups, otherwise TTN behaved as a negative APP. Apolipoprotein A-1 is a negative APP that showed significant differences in all three peptides, with Apo-A1 peptides being significantly lower than both the acute and non- acute phase groups.

Discussion

The identification and measurement of chicken serum APPs using targeted SRM was successful. It revealed that SAA is a major APP in chickens having been conclusively identified in the highly acute phase serum only. No SAA peptides were identified in non-acute phase serum, even in the

Table 1. Target peptides for each APP and the results of one way ANOVA comparing the peptide intensity for each selected peptide across the three sample groups: highly acute phase (n=3), acute phase (n=4) and a non-acute phase group (n=4).

APP	Peptide 1	Peptide 2	Peptide 3
SAA	EANYIGADK	EANYIGADKYFHAR	LDQEANEWGR
P-value	0.01	0.03	0.02
OVT	GAIEWEGIESGSVEQAVAK	VEDIWSFLSK	YFGYTGALR
P-value	0.002	0.18	0.0001
APO-A1	DAIAQFESSAVGK	LADNLDTLSAAAAK	WTEELEQYR
P-value	0.005	0.002	0.004
TTN	AADGTWQDFATGK	VEFDTSSYWK	HYTIAALLSPFSYSTTAVVSDPQE
P-value	0.04	0.051	0.29

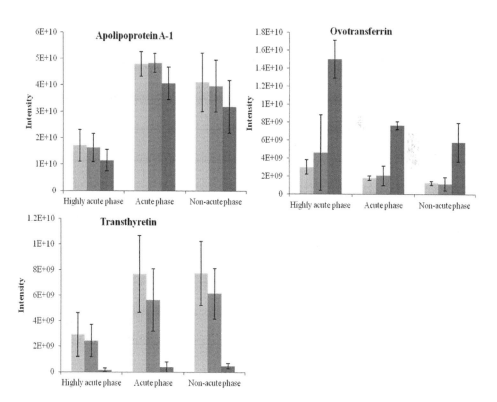

Figure 1. Graphs comparing the peptide intensities (±SD) for four of the APPs across the three groups.

fractionated lower molecular weight serum. The peptide signal for SAA was low owing to the low abundance within serum and for quantification an enrichment method is needed for normal samples. Despite the role of CRP as a major acute phase reactor in mammalian species such as pigs, humans and dogs, CRP was not identified in any sample, including the highly acute phase samples. This suggests that CRP is not an acute phase protein in chickens or that serum concentrations are lower than the limit of detection of this system. Ovotransferrin is a moderate APP in chickens and the results of this investigation corresponds well to previous work undertaken. Both Apo-A1 and TTN are known negative APPs and the results for Apo-A1 demonstrate that this APP acts in a negative fashion in the chicken also. The TTN results are less conclusive.

Taking a quantitative proteomic approach overcomes many of the issues associated with the development of immunoassays for APPs, particularly those with limited availability of high avidity antibodies. The characterisation and measurement of APPs with low serum concentrations such as SAA and CRP is also challenging, hence previous reports on these proteins in chickens are limited. Here SAA was shown to be a major APP in the chicken while revealing that CRP is unlikely to be an APP in the chicken. A study on these APPs using conventional immunoassays would have been lengthy, costly and challenging to validate, and these results highlight the usefulness of a targeted and quantifiable proteomic approach to measuring biomarkers in farm animals.

Acknowledgements

COST FAP funding for short term scientific mission (STSM). Lisette Trzpiot for all the assistance in preparing the samples. GIGA proteomic plateform, ULg.

References

Collodoro, M; Lemaire, P; Eppe, G; Bertrand, V; Dobson, R; Mazzucchelli, G; Widart, J; De Pauw, E and De Pauw-Gillet, MC (2012) Identification and quantification of concentration-dependent biomarkers in MCF-7/BOS cells exposed to 17β-estradiol by 2-D DIGE and label-free proteomics. J Proteomics 19: 4555-69

Immunoproteomic analysis of *Mycoplasma meleagridis* proteins

Ticiana S. Rocha[1]*, Alessio Soggiù[2], Maurizio Ronci[3], Luigi Bertolotti[1], Paola Roncada[2], Salvatore Catania[4], Andrea Urbani[5], Luigi Bonizzi[2] and Sergio Rosati[1]

[1]*Dipartimento di Scienze Veterinarie, Torino, Italy; ticiana.silvarocha@unito.it*
[2]*Istituto Sperimentale Italiano L. Spallanzani, Sezione di Proteomica Microbica, qualità e sicurezza degli alimenti, Dipartimento di Scienze Veterinarie e Sanità Pubblica, Università degli Studi di Milano, Milano, Italy*
[3]*Università degli Studi 'G. D'Annuzio' Chieti-Pescara, Dipartimento di Scienze Sperimentali e Cliniche, Chieti, Italy*
[4]*Laboratorio di Medicina Aviare, U.O. Micoplasmi Istituto Zooprofilattico Sperimentale delle Venezie, Legnaro, Italy*
[5]*Fondazione Santa Lucia-Istituto di Ricovero e Cura a Carattere Scientifico, Roma, Italy*

Introduction

Mycoplasma meleagridis (MM) is widespread in turkey flocks and could also establish natural infection in chickens (Mardassi *et al.*, 2007, Béjaoui Khiari *et al.*, 2011).

A number of antigens appear to be shared between MM and the more important poultry mycoplasmas, as *Mycoplasma gallisepticum* (MG) and *Mycoplasma synoviae* (MS), resulting in cross-reactivities, but there are few molecular data concerning the MM species. Although some studies have revealed antigenic heterogeneity between strains, genes encoding MM immunodominant antigens have not yet been isolated and characterized.

The aim of this work was identify and initially characterize MM proteins that could be used as potential candidates for a more reliable diagnosis by using 2D electrophoresis analysis combined with immunoenzymatic and mass spectrophotometry assays.

Materials and methods

A MM field sample, isolated from a turkey cloacal swab, identified by DGGE-PCR and immunoperoxidase staining was used. A preliminary 1D WB of whole cell proteins (5 ug protein/lane) was tested against two commercial MM Positive Controls (MM+), eight different sera from MM naturally infected turkeys, MG Positive Control (MG+), MS Positive Control (MS+) and a negative control (C-).

MM cells were prepared as described previously by Regula *et al.* (2000) and the sample was precipitated using the Methanol/Chloroform/Water Protein Precipitation. Once the protein concentration was determined by using Bradford protein assay, it was diluted to a suitable concentration for the successive separation steps.

Seven cm strips over the pH range of 3-10 (GE Healthcare) were subjected to isoelectric focusing (IEF), equilibrated and transferred onto 12% SDS-polyacrylamide gels. Of the total gels, five (protein concentration of 60 µg), were electro transferred to PVDF membranes for separately WB analyses using two different positives naturally infected serum, a C-, a MG+ and a MS+ serum. In the other two gels, total proteins were visualized or by Blue Coomassie staining or by Silver Staining.

Gel images were acquired with a calibrated scanner (ImageScanner III, GE Healthcare) at 600 dpi and 16 bit grayscale then the images were imported into Progenesis SameSpots (v4.5; Nonlinear Dynamics, UK) for analysis. All imported images were processed with QC module to check image quality (saturation, bit depth, editing). The images were then automatically analyzed using 2D analysis module for reference image selection, alignment and prefiltering. All spots were manually reviewed and validated to ensure proper detection and matching. Immunoreactive spots present only in 2D-WB probed with MM positive sera were excised and identified by mass spectrometry.

A total of 21 spots were selected, excised from the gel and subjected to in-gel digestion and peptide mass fingerprint analysis using a Maxis ESI-QUAD-TOF instrument (BrukerDaltonics, Bremen).

The MM genome was sequenced by Illumina MiSeq sequencing technology, resulting in 11 unordered contigs and using Geneious software, all possible Open Reading Frames (ORFs) from the sequence were determined.

Results

In the 2D-Western Blot analysis, several immunospots were observed in both MM positive serum, no cross reaction with the MS+ and faint cross reaction in the samples C- and MG+ of around 40 Kd. A total of 21 immunoreactive spots present only in 2D-WB probed with MM positive sera were excised and identified by mass spectrometry.

MS/MS spectra were searched against a custom database obtained from the NGS seq using Mascot 2.4 (Matrix Science), after converting the acquired MS/MS spectra to mascot generic file format. The Mascot search parameters were; 'trypsin' as enzyme allowing up to 2 missed cleavages, carbamidomethyl as fixed modification, oxidation of M as variable modifications, 5 ppm MS/MS tolerance and 0.01 Da peptide tolerance and top 20 protein entries. Homology search were performed using Blastp against NCBInr database, taxa bacteria and protein families were obtained using Pfam 27.0 (Finn *et al.*, 2014). The *in silico* analysis allowed us to identify a set of seven immunoreactives proteins.

Discussion and conclusions

The combination of high resolution protein separation by 2DE and mass spectrometry has proven to be an essential tool for proteomics (Piras *et al.*, 2012) and in our study were applied in order to identify MM proteins that could be used as potential candidates for its diagnosis.

By using Blast algorithm to find homologies with known bacterial proteins and Pfam database to find specific domains inside these proteins we were able to characterize only two of the total seven proteins initially identified, due to difficulties to have good annotation for the custom database.

The first protein belongs to the family of lipoprotein X, being found along with a C-terminal domain in a group of Mycoplasma lipoproteins of unknown function, but also matching the well characterized P80 lipoprotein from *Mycoplasma agalactiae* (MA), which is an important antigen for the MA diagnostic.

The second protein belongs to the family of Ornithine carbamoyltransferase, which converts citrulline in the presence of phosphate to ornithine and carbamoylphosphate in nonglycolytic mycoplasmas. Its application for the detection of *Mycobacterium bovis* in cattle serum was recently demonstrated (Wei *et al.*, 2014).

More importantly, both proteins have a low similarity with the others avian Mycoplasmas and seems to represent good candidates for a MM serological tests.

References

Béjaoui Khiari, A., A. Landoulsi, H. Aissa, B. Mlik, F. Amouna, a Ejlassi, *et al.*, Isolation of Mycoplasma meleagridis from chickens., Avian Dis. 55 (2011) 8-12.

Finn, R.D., A. Bateman, J. Clements, P. Coggill, R.Y. Eberhardt, S.R. Eddy, *et al.*, Pfam: the protein families database., Nucleic Acids Res. 42 (2014) D222-30. http://dx.doi.org/10.1093/nar/gkt1223.

Mardassi, B.B.A., A Béjaoui Khiari, L. Oussaief, A Landoulsi, C. Brik, B. Mlik, *et al.*, Molecular cloning of a Mycoplasma meleagridis-specific antigenic domain endowed with a serodiagnostic potential., Vet. Microbiol. 119 (2007) 31-41. http://dx.doi.org/10.1016/j.vetmic.2006.08.008.

Piras, C., A. Soggiu, L. Bonizzi, A. Gaviraghi, L. Deriu, L. De Martino, *et al.*, Comparative proteomics to evaluate multi drug resistance in Escherichia coli., Mol. Biosyst. 8 (2012) 1060-7. doi:10.1039/c1mb05385j.

Regula, J.T., B. Ueberle, G. Boguth, A. Görg, M. Schnölzer, R. Herrmann, *et al.*, Towards a two-dimensional proteome map of Mycoplasma pneumoniae Proteomics and 2-DE, Electrophoresis. 21 (2000) 3765-3780.

Wei, Z., Z. Xiangmei, L. Yun, P. Yun, L. Zhu, L. Jingjun, Y. Lifeng, *et al.*, Mycobacterium Bovis Ornithine Carbamoyltransferase, MB1684, Induces Proinflammatory Cytokine Gene Expression by Activating NF-κB in Macrophages., DNA and Cell Biology. 33 (2014) 311-319. http://dx.doi.org/10.1089/dna.2013.2026.

Comparative analysis of acute phase proteins in farm animals by affinity methods coupled to proteomics

Lourdes Soler[1]*, Fermín Lampreave[1], M.A. Álava[1], Richard J.S. Burchmore[2] and Peter D. Eckersall[3]
[1]Departamento de Bioquímica y Biología Molecular y Celular. Facultad de Ciencias. Universidad de Zaragoza, Pedro Cerbuna 12, 50009, Zaragoza, Spain; lourdes_soler_baigorri@hotmail.com
[2]Glasgow Polyomics, University of Glasgow, Bearsden Rd, Glasgow, G61 1QH, Scotland, United Kingdom
[3]Institute of Biodiversity, Animal Health and Comparative Medicine, University of Glasgow, Bearsden Rd, Glasgow, G61 1QH, Scotland, United Kingdom

Introduction

Farm animal sciences are essentially aimed at understanding biological mechanism related to the production of food. According to the Food and Agriculture Organization (FAO) of the United Nation, pork is the most consumed meat from terrestrial animals and development of pig production is related with increases in animal health. Proteomics studies in farm animals have focused on monitoring and enhancing animal health and welfare through the search of relevant biomarkers (Bendixen *et al.*, 2011). Acute phase proteins (APP) are serum proteins whose concentration changes after infection, inflammation or trauma, which can be useful in disease diagnosis, and for this reason APPs have been proposed as biomarkers of disease. It has been shown that lipoprotein metabolism is altered by inflammation (Khovidhunkit *et al.*, 2004), also the role of lipoproteins as important mediators of the immune response and host defense mechanism has been demonstrated (Chait *et al.*, 2005; Getz 2005). The aim of this study was to pre-fractionate pig serum by lipid precipitation and by heme affinity, for characterization of low abundance protein in concentration during an acute phase reaction.

Materials and methods

Selected normal pig sera which contained a CRP concentration below to 0.1 mg/ml were selected. Pig serum samples with CRP above 0.1 mg/ml were considered acute phase sera. Calcium silicate hydrate (CSH) (Sigma Poole, UK) was used as lipid precipitation reagent, 20 µl of CSH (100 mg/ml 50 mM ammonium bicarbonate) were mixed with different serum volumes. Lipid precipitation was made in whole pig sera and in serum fractions obtained after gel filtration chromatography on a Superdex 200 10/300 GL column (GE Healthcare, UK). Cholesterol levels were measured by routine clinical laboratory methods in samples before and after lipid precipitation, and protein concentration after gel chromatography was determined using Bradford reagent. Hemin-agarose Bovine (Type I) (Sigma Poole, UK) was used for the adsorption of heme-binding proteins. Serum samples (25 µl) were incubated with different volumes of Hemin-agarose. Samples from each treatment were run on SDS-PAGE gels (Bio-Rad) and stained with Coomassie blue. Selected bands were excised, digested in-gel with trypsin and analysed by LC-MS/MS.

Results

Treatment 2 was selected as the best treatment with CSH to remove cholesterol and separate proteins by SDS-PAGE (Figure 1). Proteins between 130 to 70 KDa have been concentrated in pellets compared with whole sera. Most of proteins identified were apolipoproteins (bands 1, 13, 14, 16 and 18) and complement components (bands 4, 6, 7, 15 and 13'). PigMAP (major acute phase protein in pigs) was identified in band 5 and Apo A-I (negative acute phase protein) identified in band 16, showed the largest change in abundance (Gonzalez-Ramon *et al.*, 1995). One fragment of complement C3 precursor (band 13') in the AP serum was not found in healthy sera, and is proposed as a potential biomarker. Fractionation of pig sera according to lipid composition was made by gel filtration chromatography (Gordon *et al.*, 2013). According to the gel filtration chromatogram, along with the cholesterol and protein concentration the LDL lipoproteins were in fractions 8, 9 and 10 (Figure 2). HDL lipoproteins were in different fractions, comparing healthy serum (12, 13, 14 and 15) with diseased serum (14, 15, 16, and 17). After lipid precipitation, it was noted that LDL fraction band 2 showed the largest increase in acute phase response. The LDL fraction also showed a significant increase in band 9 (α2 macroglobulin), and APPs such as PigMAP (band 5) and Apo A-I (band 10) were in the HDL fraction. Contrary to anticipated results hemopexin wasn't found in hemin-agarose affinity bound fractions (Figure 3). However, some of proteins bounded to hemin-agarose also were associated to lipids. These proteins were: complement factor H (band 4 Figure 1, band 1 Figure 3), PigMAP (band 5 Figure 1, band 2 Figure 3), gelsolin (band 8 Figure 1, band 3 Figure 3), IgG heavy chain (band 12 Figure 1, band 6 Figure 3) and Apo A-I (band 16 Figure 1, bands 7, 8 Figure 3).

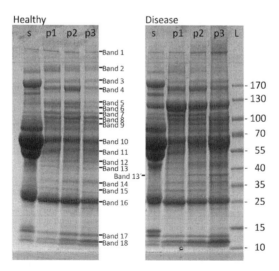

Figure 1. Electrophoresis from pig serum (s), and pellets (p) after treatment (1,2,3) with CSH. Selected bands from a healthy pig (lane p2, bands 1-18) and from a diseased pig (lane p2, band 13´) were sent to LC-MS/MS.

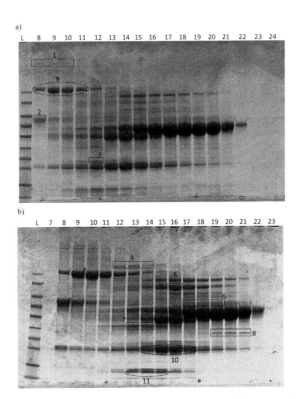

Figure 2. Pellet after lipid precipitation from each fraction after gel filtration chromatography from a) healthy serum and b) diseased serum. Bands 1-5 sent to LC-MS/MS.

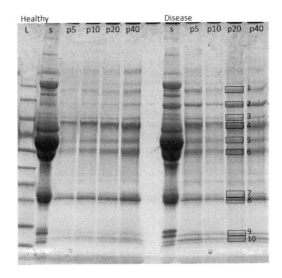

Figure 3. Affinity chromatography of heme-binding proteins in pig serum samples; (L) ladder, (s) whole serum, (p) pellet obtained after incubating 25 μl serum with different amounts of Hemin agarose.

Conclusion

Pre-fractionation of pig sera allowed a focus on proteins that in whole serum are in low concentration. This approach is a useful tool in the identification of novel biomarkers during an acute phase response. It is interesting to underline the presence of common proteins in both treatments (CSH and hemin-agarose). Further studies will be necessary, but these results could suggest interactions between these proteins.

Acknowledge

Thanks to FA1002 COST Action Management Committe for financial support.

References

Bendixen, E., M. Danielsen, K. Hollung, E. Gianazza and I. Miller (2011). Farm animal proteomics--a review. J Proteomics 74(3): 282-293.

Chait, A., C. Y. Han, J. F. Oram and J. W. Heinecke (2005). Thematic review series: The immune system and atherogenesis. Lipoprotein-associated inflammatory proteins: markers or mediators of cardiovascular disease? J Lipid Res 46(3): 389-403.

Getz, G. S. (2005). Thematic review series: the immune system and atherogenesis. Immune function in atherogenesis. J Lipid Res 46(1): 1-10.

Gonzalez-Ramon, N., M. A. Alava, J. A. Sarsa, M. Pineiro, A. Escartin, A. Garcia-Gil, F. Lampreave and A. Pineiro (1995). The major acute phase serum protein in pigs is homologous to human plasma kallikrein sensitive PK-120. FEBS Lett 371(3): 227-230.

Gordon, S. M., J. Deng, A. B. Tomann, A. S. Shah, L. J. Lu and W. S. Davidson (2013). Multi-dimensional co-separation analysis reveals protein-protein interactions defining plasma lipoprotein subspecies. Mol Cell Proteomics 12(11): 3123-3134.

Khovidhunkit, W., M. S. Kim, R. A. Memon, J. K. Shigenaga, A. H. Moser, K. R. Feingold and C. Grunfeld (2004). Effects of infection and inflammation on lipid and lipoprotein metabolism: mechanisms and consequences to the host. J Lipid Res 45(7): 1169-1196.

An integrative study of the early immune response against ETEC

Laura Soler[1,2,3,4,5*], Marcel Hulst[6], Jan van der Meulen[6], Gabriel Mazzucchelli[7], Mari Smits[6], Edwin de Pauw[7] and Theo Niewold[5]

[1]INRA, UMR85 Physiologie de la Reproduction et des Comportements, Nouzilly, France; lsolervasco@tours.inra.fr
[2]CNRS, UMR7247, Nouzilly, France
[3]Université François Rabelais de Tours, Tours, France
[4]IFCE, Institut Français du Cheval et de l'Equitation, 37380 Nouzilly, France
[5]Division of Livestock-Nutrition-Quality, Faculty of Bioscience Engineering, K.U. Leuven, Belgium
[6]Animal Breeding and Genomics Centre, Wageningen University and Research Centre, the Netherlands
[7]Laboratory of Mass Spectrometry-GIGA-Proteomics, University of Liège, Belgium

Objectives

Enterotoxigenic *Escherichia coli* (ETEC) is an important enteric pathogen in pigs, and it is responsible for substantial economic losses every year. The molecular mechanisms that lead to ETEC colonization and infection are complex, and those decisive for the development of diarrhea and/or the different clinical presentations are still undefined. Systems biology approaches like transcriptomics and proteomics are the most adequate to investigate the whole set of cellular responses to a challenge. The transcriptional profile of the intestinal mucosa has been established in response to different ETEC infection models, showing some features of the specific activation of the local innate immune response. However, little is known on the proteomic changes induced and therefore the cellular mechanisms not directly reflected by transcriptomics remain unknown, such as those governed by transcriptional regulatory factors, post-transcriptional and post-translational modifications, protein interactions and protein abundance changes. Here we combined RNA microarray and label-free LC-MS/MS technologies to describe the transcriptional regulation of the set of differentially expressed proteins involved in the early stages of ETEC infection in the pig small intestine.

Material and methods

Mucosal scrapings were recovered from a previous study (Niewold *et al.*, 2005). In brief, four animals were subjected to the small intestine segmentary perfusion (SISP) technique, which consists in independently exposing small intestine loops of anesthetized animals to different treatments under continuous perfusion, thus allowing for isogenic comparison. Five ml of 10^9 cfu/ml of ETEC in PBS or PBS alone were inoculated per loop during one hour followed by 8 h perfusion. Mucosal scrapings were taken at hours 0 and 8, and animals were humanely euthanized thereafter. Samples were snap-frozen and kept at -80 °C until analysis (Niewold *et al.*, 2005). RNA extraction and gene expression (microarray's) analysis was performed as described elsewhere (Hulst *et al.*, 2013). Protein extraction and label free UPLC/HDMS^E proteomic analysis was performed as described elsewhere (Soler *et al.*, 2013). Significantly regulated transcripts/proteins ($P<0.05$; fold-change ±1.5) were listed, re-annotated when necessary and renamed to Hugo gene symbols. A complete systems biology analysis was performed with both datasets separately and in a third dataset obtained after joining significantly regulated transcripts and proteins in the same list. The Database for Annotation, Visualization and Integrated Discovery (DAVID version 6.7) website and the 'Set Distiller' module of GeneDecks website were employed. Common processes were identified and subsets of transcripts/proteins were elaborated according to these for interpretation of the underlying biological processes.

Results and discussion

We observed that 350 and 154 unique transcripts and proteins were respectively regulated in the ETEC-challenged small intestine segments compared to mock-challenged loops. In brief, the main processes identified based only on the differential RNA expression were related to the innate immune response, gluconeogenesis, and vascular repair/development, while those based only on the differential protein expression were related to the local antibacterial reaction, complement activation, different pathways associated with the coagulation cascade regulation as well as enhanced vasodilation. Common processes to both RNA and protein differential expression were related with PPARα signaling, arginine/proline metabolism and N-linked glycosylation, along with signaling pathways related with cell survival and cell-cell recognition.

Through the combination of the results obtained by transcriptomics and proteomics we were able to better interpret which are the main molecular events taking place during the early response of the intestinal mucosa towards ETEC. In summary, the enriched pathways/networks identified after our analysis describe the local antimicrobial response of epithelial cells, the metabolic changes associated with the initiation of the immune response, the early activation, degranulation and recruitment of neutrophils in the intestinal mucosa, the initiation of macrophage plaques and the regulation of the vascular events that allow/derive from the latter two.

Although the time-lapse and regulation existing between RNA and protein expression should be taken into account when interpreting these two datasets together, we were able to connect RNA and protein expression for certain relevant biological events. At the same time, protein expression could be validated by microarray results, given that the direction of regulation of identical or similar transcripts/proteins was generally correlated. In conclusion, a more holistic view of the early response of the small intestine towards ETEC could be obtained by combining results from high-throughput RNA and protein expression.

Acknowledgements

Laura Soler acknowledges financial support by the Martín Alonso Escudero Foundation.

References

Hulst, M., Smits, M., Vastenhouw, S., de Wit, A., Niewold, T. and van der Meulen, J., 2013. Transcription networks responsible for early regulation of *Salmonella*-induced inflammation in the jejunum of pigs. J Inflamm (Lond) 10: 18.

Niewold, T.A., Kerstens, H.H., van der Meulen, J., Smits, M.A. and Hulst, M.M., 2005. Development of a porcine small intestinal cDNA micro-array: characterization and functional analysis of the response to enterotoxigenic *E. coli*. Vet Immunol Immunopathol 105: 317-329.

Soler, L., Niewold T., De Pauw, E. and Mazzucchelli, G., 2013. Small intestinal response to enterotoxigenic *Escherichia coli* infection on pigs as revealed by label free UPLC/MSE proteomics. 3rd meeting of working groups 1, 2 and 3 of COST Action FA1002., Kosice, Slovakia. Wageningen Academic Publishers, pp. 55-58.

Effect of the use of OTC as feed additive in the pig serum proteome

Laura Soler[1,2], Ingrid Miller[3], Karin Hummel[4], Flemming Jessen[5], Manfred Gemeiner[3], Ebrahim Razzazi-Fazeli[4] and Theo Niewold[2]*

[1]*Livestock-Nutrition-Quality Division, Biosystems Department, Faculty of Biosciences Engineering, KU Leuven, Kasteelpark Aremberg 30, Heverlee 3001, Belgium*

[2]*INRA, UMR85 Physiologie de la Reproduction et des Comportements, 37380 Nouzilly, France; lsolervasco@tours.inra.fr*

[3]*Department of Biomedical Sciences, University of Veterinary Medicine, 1210 Vienna, Veterinaerplatz 1, Austria*

[4]*VetCore, University of Veterinary Medicine, 1210 Vienna, Veterinaerplatz 1, Austria*

[5]*National Food Institute, Technical University of Denmark, Søltofts Plads, 2800 Kgs. Lyngby, Denmark*

Objectives

Subtherapeutical concentrations of antibiotics such as oxytetracycline (OTC) have been included as growth promoters in the feed of intensively bred animals for years, yet their use is banned nowadays. Although their effect was traditionally attributed to their role in modulating the intestinal microbiota, recent studies suggest that they might exert a more important anti-inflammatory role. In fact, they might have an anabolic effect through the tuning of metabolic inflammation, which is enhanced in intensively bred animals receiving high energy meals (Khadem *et al.*, 2014). The discovery of the exact mechanism of action of antibiotics when used as growth promoters is very important in the search of natural alternatives. Our objective was to investigate the systemic effects of the subtherapeutic OTC administration in pigs through the identification of the serum proteomic changes induced by such treatment.

Material and methods

Thirty four-week old piglets were divided in two groups. The experimental period lasted 37 days in which control piglets (n=15) received only a commercial diet, whereas treated animals (n=15) received 250 ppm of commercial 80% OTC mixed with feed. Piglets were weighed at days 0, 32 and 65 of the test period. At test day 37, blood was collected by jugular venipuncture using single-use blood collection tubes without any additive. Blood was allowed to clot for 1 hour at room temperature and then centrifuged at 2,000×*g* for 15 min to obtain serum. The acute phase response markers haptoglobin and serum amyloid A were determined using commercial kits (Tridelta, Ireland). From all serum samples collected, four of each group were randomly selected from control and OTC-supplemented piglets and subjected to two-dimensional differential gel electrophoresis (Miller, 2012), both in non-fractionated form and after a depletion/enrichment step. Three different depletion/enrichment methods were employed: protein enrichment on hexapeptide resin (Proteominer Kit, BioRad, Hercules, CA, USA), albumin and IgG depletion (ProteaPrep

Kit, Protea, Morgantown, USA) and hydrophobic protein enrichment through in-house Triton X cloud point separation (Flemming *et al.*, unpublished results). Differentially expressed spots ($P<0.05$; Fold-change ± 1.3) were identified by two-way ANOVA using DeCyder 7.0 software (GE Healthcare Lifesciences, Munich, Germany), excised from the gel and subjected to MALDI-TOF/TOF (Ultraflex II, Bruker Daltonics, Bremen, Germany) analysis for protein identification. Processed spectra were searched via an in-house Mascot server (Matrix Science, Boston, MA, USA) in the NCBI database using the following search parameters: taxonomy: mammals; global modifications carbamidomethylation on cysteine; variable modifications: Oxidation (M); MS tolerance 100 ppm; MS/MS tolerance 1 Da; one missed cleavage allowed. Identifications were considered statistically significant where $P<0.05$. Regulated proteins were listed as official HUGO gene names and subjected to systems biology analysis using DAVID and GeneDecks Set Distiller module. Proteomics results were confirmed/validated by ELISA (Total IgG determination USCN, Hubei, China), immunoblotting (alpha-1 acid glycoprotein, in-house antibody; Dilda *et al.*, unpublished results) and colorimetric assays (ferric reducing anti-oxidant power to measure total serum antioxidant capacity and an adapted p-nitrophenyl acetate assay to determine serum paraoxonase-1 protein; Escribano *et al.*, unpublished results). Differences in values of each parameter described above were analyzed by two-tailed Student's t-tests using GraphPad Prism 5 for Windows (GraphPad software, La Jolla, USA). The significance level was set at $P<0.05$ in all cases.

Results and discussion

The effect of the supplementation of animal feed with OTC at sub-therapeutically concentrations was reflected in a significantly higher weight gain of supplemented animals at the end of the experimental period compared to controls ($P<0.05$). In the same animals, OTC supplementation also produced (not significantly) lower levels of the inflammation markers serum haptoglobin and significantly lower levels of serum amyloid A ($P<0.05$). Combining results from whole and enriched serum analysis, differences in spot intensity between gels from control and OTC-supplemented animals were identified in 81 matched different spots ($P<0.05$). In total, thirteen different proteins were identified in thirty-one different spots. For the remaining spots the amount of protein was insufficient to confidently identify the contained proteins by MALDI-TOF/TOF). From the regulated proteins, those that were confidently identified were analyzed in DAVID and GeneDecks to detect significantly enriched pathways and GO biological functions. Different pathways and GO terms related with the innate immune response, antioxidant activity and lipid metabolism were called significant. We also observed that the significantly enriched transcription binding factor sites within the regulated proteins were all related to the metabolic switch that occurs during inflammation from anabolic to catabolic pathways. In all, results indicated that OTC supplementation produced changes in the serum proteome associated with the reduction of metabolic inflammation. Systems biology analysis of proteomic results allowed us to link the systemic inflammatory condition, metabolic status and growth rates in intensively produced pigs, and how they are implemented with OTC supplementation. The obtained results support the 'immunomodulatory theory' as the main mechanism of action of antimicrobial growth promoters.

Acknowledgements

Laura Soler acknowledges financial support by the Martín Alonso Escudero Foundation. D. Escribano, J.J. Cerón and K. De Backer are acknowledged for their help during results validation.

References

Khadem, A., Soler, L., Everaert, N. and Niewold, T.A., 2014. Growth promotion in broilers by both oxytetracycline and Macleaya cordata extract is based on their anti-inflammatory properties. Br J Nutr: 1-9.

Miller, I., 2012. Application of 2D DIGE in animal proteomics. Methods Mol Biol 854: 373-396.

Total alkaline phosphatase and bone alkaline phosphatase in dairy cows during periparturient period

Jože Starič*, Marija Nemec and Jožica Ježek
Clinic for ruminants, Veterinary faculty, University of Ljubljana, Gerbičev 60, Ljubljana, Slovenia;
joze.staric@vf.uni-lj.si

Objectives

Alkaline phosphatase (ALP) forms a large family of dimeric enzymes, usually confined to the cell surface, which hydrolyze various monophosphate esters at a high pH optimum with release of inorganic phosphate. Mammalian ALP are metalloenzymes encoded by a multigene family. Three metal ions including two Zn^{2+} and one Mg^{2+} in the active site are essential for enzymatic activity. In mammalian ALP there is also Ca^{2+} necessary for normal function. Slight differences in electrophoretic mobility and thermo stability between ALP from various tissues are attributed to slight differences in amino acids structure and post-translational modification (Sharma *et al.*, 2014).

Bone alkaline phosphatase (bALP) is the most promising marker of bone formation in human medicine (Christenson, 1997). Its role is in mineralization of bone tissue. In adult humans it contributes about half of ALP activity (Sharma *et al.*, 2014).

High producing dairy cows are unique for very dramatic metabolic changes when they go from dry period into intensive milk production. This period is called also transition period. It starts about 3 weeks before calving and lasts until about 3 weeks after calving and is the most critical time in a dairy cows production cycle. Since milk production in dairy cows is unnaturally high, a cow has to dramatically adapt many metabolic activities to sustain homeostasis. Among the most charged organs in this process is liver, which is responsible for energy and protein supply to the cow and also bones as the major depot of Ca. Many cows do not manage this metabolic strain and acquire metabolic diseases, especially hypocalcaemia and ketosis, which have negative effects on health, welfare, reproduction and production. ALPs are from liver and bones and could indicate these changes. There are still many gaps in the knowledge about dynamics and role of ALP and bALP in dairy cattle during transition period.

The aim of this study is to examine dynamics of ALP (predominantly of liver and bone origin) and bALP (just bone origin) during the transition period in dairy cows.

Material and methods

The study was conducted at a dairy farm with intensive milk production, with loose housing system on cubicles and pasture during warm part of the year. Forty-one clinically healthy cows before at least the 4th lactation were included in the study, 20 during winter time and 21 during summer time. During winter the animals were indoors all the time, while during summer they were also on pasture every day. Cows were fed usual winter ration based on home produced forages: grass silage, maize silage, hay and straw according to NRC (2001) recommendations. No anion salts were added to the diet and DCAD of the ration ranged from +200 to +300 mEq/kg dry matter. We monitored and obtained blood samples from investigated animals four times during the transition period:

- 1 month before calving;
- 10 or less days before calving;
- within 48 hours after calving; and
- 10 to 20 days in milk.

Venous blood samples were collected in vacutainer tubes (10 ml) with no additive from v. caudalis mediana according to the protocol between 9 and 11 a.m., to avoid daily fluctuations in analytes. Harvested blood serum was stored at -20 °C until analyses.

Blood serum ALP activity was measured with automatized biochemical analyzer RX Daytona (Randox, Ireland) according to manufacturer's instructions. bALP activity was measured using Alkphase-B kit (Metra Biosystems, USA) by EIA according to manufacturers' instructions. The absorbance of coloured end product was measured by optical reader at 405 nm.

Obtained values were statistically analyzed using descriptive statistics and correlations (Pearsons'). All the values that did not distribute normally were normalized. The influence of season and time of blood sampling, on our dependent variables was tested using analysis of variance with repetition. If the factor value was not distributing normally nonparametric Fridman's test was used. Multiple comparisons as 'post hoc' analysis were performed using Bonferroni's correction of p-value. Statistical significance was set at $P<0.05$.

Results and discussion

Season did not influence blood levels of ALP while time of sampling did during the summer ($F(1.67, 65.19)=10.99$, $P<0.001$) and winter ($\chi^2(3)=24.02$, $P<0.001$) (Table 1). The same applies also for bALP, where just time of sampling statistically significantly influenced values in summer ($F(2.24, 87.22)=49.24$, $P<0.001$) and in winter ($F(3, 57)=28.35$, $P<0.001$) group. We expected that values of bALP would be higher during summer when animals had also access to pasture and more physical activity (Allen, 2003).

In the summer group ALP was statistically significantly higher in 3. sampling than in other three ($P<0.05$). Further the average value of ALP in 2. sampling was statistically significantly higher than in 1. ($P<0.05$). In the winter group ALP was statistically significantly lower in samplings 4 than in 2 (before calving) and 3 ($P<0.05$).

Values of bALP in the summer group were statistically significantly lower in 1. and 2. sampling than in samplings 3 and 4 (after calving) ($P<0.05$). In winter group it is statistically significantly

Table 1. Serum levels of ALP and bALP in dairy cows in four samplings during winter and summer time.

Sampling	ALP (U/l)		bALP (U/l)	
	Summer	Winter	Summer	Winter
1	39.76±13.25[2,3]	45.05±22.18	12.10±3.38[3,4]	15.38±4.36[3,4]
2	44.22±13.44[1,3]	47.85±22.43[4]	15.09±7.69[3,4]	16.96±9.01[3,4]
3	51.85±15.47[1,2,4]	50.50±15.12[4]	25.71±10.15[1,2]	31.02±7.22[1,2,4]
4	38.80±12.66[3]	36.35±10.95[2,3]	29.23±7.26[1,2]	24.26±5.11[1,2,3]

[1,2,3,4] Indicators of statistically significant differences.

higher ($P<0.05$) in sampling 3 compared to other three samplings. Further also in the sampling 4, the values were statistically significantly higher ($P<0.05$) than in samplings 1 and 2.

The association between ALP and bALP was tested on joined summer and winter groups of cows due to insignificant difference of values between summer and winter groups. We established a highly statistically significant positive correlation between ALP and bALP ($P<0.001$). Total ALP correlates well with bALP and can be used in assessing bone metabolism, given that there is no liver pathology present.

High bALP activity post partum can be associated with more intensive bone metabolism that provides Ca for the steep increase in demand (Goff, 2000; Starič and Zadnik, 2010). Apparently anabolic bone metabolism is possibly important also for strengthening of weekend connections between the pelvic bones after delivery of a calf.

Acknowledgements

This work was financially supported by the Slovenian Research Agency; program P4-0092 (Animal health, environment and food safety).

References

Allen, M.J., 2003. Biochemical markers of bone metabolism in animals: uses and limitations. Veterinary Clinical Pathology 32, 101-113.

Christenson, R.H., 1997. Biochemical markers of bone metabolism: an overview. Clinical Biochemistry 30, 573-593.

Goff, J.P., 2000. Pathophysiology of calcium and phosphorus disorders. Veterinary Clinic of North America, Food Animal Practice 16, 319-337.

National research council, 2001. Nutrient requirements of dairy cattle. 7th ed. National Academy Press, Washington, USA, pp. 258-80.

Sharma, U., Pal, D., Prasad, R., 2014. Alkaline Phosphatase: An Overview. Indian Journal of Clinical Biochemistry 29(3), 269-278.

Starič, J. and Zadnik, T., 2010. Biochemical markers of bone metabolism in dairy cows with milk fever. Acta Veterinaria (Beogr.) 60, 401-410.

Genomics and deep proteome profiling of *Staphylococcus epidermidis* for uncovering adaptation and virulence mechanisms

Pekka Varmanen[1]*, Pia Siljamäki[1,2], Tuula A. Nyman[2] and Kirsi Savijoki[1]
[1]*Department of Food and Environmental Sciences, University of Helsinki, Finland;*
pekka.varmanen@helsinki.fi
[2]*Institute of Biotechnology, University of Helsinki, Finland*

Introduction

Staphylococcus epidermidis (SE) is an opportunistic pathogen capable of infecting humans and animals. It is one of the major pathogens associated with hospital-acquired infections as well as subclinical intramammary infections (IMIs) in dairy cows. In humans, SE is part of the normal and balanced skin microbiota, which may block the colonization of potentially harmful microbes such as *Staphylococcus aureus*. On the other hand, epidemiological studies suggest that humans could be an important source of SE-mediated bovine IMIs. Our recent findings support this by showing that the genome of an SE strain (PM221) isolated from subclinical bovine IMI resembles more closely the commensal-type human ATCC12228 strain than the highly virulent sepsis-associated strain RP62A (Figure 1). The bovine strain shares also some genetic factors (e.g. ACME element, genes for Aap, haemolysins, proteases and lipases) common with the virulent RP62A strain as well as the marker gene associated with commensal-type human SE strains (Savijoki *et al.,* 2014). The marker gene found in the PM221 genome was fdh (formate dehydrogenase) which can be used to distinguish between non-pathogenic/commensal and pathogenic-type human SE strains (Conlan and Mijares *et al.,* 2012). However, our findings suggest that this marker gene may not be applicable to discriminate between commensal and infectious bovine strains. Other distinct features of the PM221 genome was the vast number of gene paralogs and three prophage-type elements, which may have provided the bovine strain with better tools to survive and adapt in the bovine host (Savijoki *et al.,* 2014).

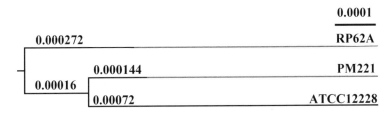

Figure 1. PHYML analysis of one-to-one protein orthologs (known ribosomal proteins) from PM221, ATCC12228 and RP62A with branch lengths representing evolutionary times. Distance scale is shown above the plot.

Results

To complement the genome-level findings and to gain deeper insight into the adaptation and virulence mechanisms of SE we applied 2D-DIGE, cell-surface shaving proteomics, total proteome and exoproteome cataloging using GeLC-MS/MS for comparing the proteomes between the bovine (PM221) and human (ATCC12228) strains. The 2D-DIGE analyses shed light on the proteome dynamics in different growth phases and pointed out high similarity between PM221 and ATCC12228 (Figure 2). Further analyses revealed that PM221 shares similarities with both RP62A and ATCC12228; PM221 and RP62A are likely to exploit biofilm formation for adaptation, whereas PM221 and ATCC12228 down-regulate the TCA (tricarboxylic acid) activity and stimulate the formation of small colony variants for increasing stationary phase survival and persistence (Savijoki *et al.*, 2014). Although the human ATCC12228 strain was able to infect a bovine host, the PM221 strain caused more severe clinical signs. Surfacome-shaving proteomics revealed strain- and condition-specific differences associated with certain adhesive moonlighting proteins and the bona fide surface proteins, which may explain the ability of the ATCC12228 strain to induce persisting bovine IMI and the differences in the clinical effects of the human and bovine strains.

The use of GeLC-MS/MS applied to total proteome (Siljamäki *et al.*, 2014a) and exoproteome analysis (Siljamäki *et al.*, 2014b) increased substantially the number of identifications and revealed that PM221 and RP62A express wider range of different virulence factors than ATCC12228. Interestingly, a large number of the identifications was associated with cytoplasmic proteins, many of which have previously been identified from membrane vesicle (MV) fractions produced by *Staphylococcus aureus*. This non-classical secretion pathway is known to enable protein export in a protected and concentrated manner to reach their final destinations in the host cells or other

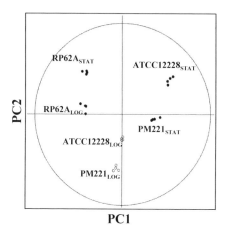

Figure 2. PCA analysis shows the clustering of different DIGE spot maps by two principle components. The PM221 and ATCC12228 groups cluster closer together, whereas those of RP62A form a separate cluster from the other four experimental groups.

bacterial cells (Mashburn-Warren and Whiteley, 2005). For testing the possibility of MV-mediated protein export in SE, the culture supernatants from the human and bovine strains were subjected to ultrafiltration and quantitative 1DE analysis, which revealed specific enrichment of small molecular weight (<100 kDa) proteins. This suggests that some of the cytoplasmic proteins might have been exported out of the cell in large molecular weight structures like MVs (Siljamäki *et al.*, 2014b).

Conclusions

Our findings highlight the ability of SE to exploit diverse strategies, involving biofilm formation, specific changes associated with late stationary phase metabolism and non-conventional pathway for protein export, which all aim to ensure better adaptation and/or successful infection. Our findings also strengthen the hypothesis that humans could be an important source of SE-mediated infections in dairy cows.

Acknowledgements

This study was supported by the Academy of Finland (grants 139296 to P. Varmanen and 135628 and 140950 to T. Nyman), and the Walter Ehrström Foundation to K. Savijoki.

References

Conlan, S., Mijares, L.A., NISC Comparative Sequencing Program, Becker, J., Blakesley, R.W., Bouffard, G.G., Brooks, S., Coleman, H., Gupta, J., Gurson, N., Park, M., Schmidt, B., Thomas, P.J., Otto, M., Kong, H.H., Murray, P.R., and Segre, J.A., 2012, *Staphylococcus epidermidis* pan-genome sequence analysis reveals diversity of skin commensal and hospital infectionassociated isolates. Genome Biology, 13, R64.

Lee, E.Y., Choi. D.Y., Kim, D.K., Kim, J.W., Park, J.O., Kim, S., Kim, S.H., Desiderio, D.M., Kim, Y.K., Kim, K.P., and Gho, Y.S., 2009, Gram-positive bacteria produce membrane vesicles: Proteomics-based characterization of *Staphylococcus aureus*-derived membrane vesicles. Proteomics 9, 5425-5436.

Mashburn-Warren, L.M., and Whiteley, M., 2005, Membrane vesicle traffic signals and facilitates group activities in a prokaryote. Nature 437, 422-425.

Savijoki, K,, Iivanainen, A., Siljamäki, P., Laine. P.K., Paulin. L., Karonen. T., Pyörälä. S., Kankainen. M., Nyman. T.A., Salomäki. T., Koskinen. P., Holm. L., Simojoki. H., Taponen. S., Sukura. A., Kalkkinen. N., Auvinen. P., and Varmanen, P., 2014, Genomics and proteomics provide new insight into the commensal and pathogenic lifestyles of bovine- and human-associated *Staphylococcus epidermidis* strains. Journal of Proteome Research 13, 3748-3762.

Siljamäki, P., Varmanen, P., Kankainen, M., Pyörälä, S., Karonen, T., Iivanainen, A., Auvinen, P., Paulin, L., Laine, P.K., Taponen, S., Simojoki, H., Sukura, A., Nyman, T.A., and Savijoki, K., 2014a, Comparative proteome profiling of bovine and human *Staphylococcus epidermidis* strains for screening specifically expressed virulence and adaptation proteins. Proteomics 14, 1890-1894.

Siljamäki, P., Varmanen, P., Kankainen, M., Sukura, A., Savijoki, K., Nyman, T.A., 2014b, Comparative exoprotein profiling of different *Staphylococcus epidermidis* strains reveals potential link between nonclassical protein export and virulence. Journal of Proteome Research 13(7):3249-3261.

Blood fibrinogen concentration in New Zealand White Rabbits during first three months of their life

Evgenya V. Dishlyanova[1]*, Teodora M. Georgieva[1], Vladimir S. Petrov[2], Tatyana Vlaykova[3], Fabrizio Ceciliani[4], Radina N. Vasileva[5] and Ivan P. Georgiev[1]

[1]Department of Pharmacology, Animal Physiology and Physiological Chemistry, Trakia University, 6000 Stara Zagora, Bulgaria; dishlianova@yahoo.com

[2]Department of Veterinary Medical Microbiology, Infection and Parasitic Diseases, Faculty of Veterinary Medicine, Trakia University, 6000 Stara Zagora, Bulgaria

[3]Department of Chemistry and Biochemistry, Faculty of Medicine, Trakia University, 6000 Stara Zagora, Bulgaria

[4]Dipartimento di Scienze Animali e Salute Pubblica, Università degli Studi di Milano, Milano, Italy

[5]Student from Faculty of Veterinary Medicine, Trakia University, 6000 Stara Zagora, Bulgaria

Introduction

The term acute phase protein refers to the inflammatory response of the host occurring shortly after tissue injury. Kushner at al. 2006 (Kushner *et al.*, 2006), classified positive acute phase proteins (APP) into 3 groups: (1) APP whose concentration increased by 50% (ceruloplasmin and C_3); (2) APP exhibiting 2- to 3-fold increase (haptoglobin, fibrinogen and α-albumins with antiprotease activity); and (3) proteins, increasing extremely rapidly up to 1000 times (C-reactive protein and SAA). In fact, 98% of plasma (serum) protein content consists of 22 proteins, and the remaining 2% comprise about 1000 proteins, which could serve as biomarkers (Stanley and Van Eyk, 2005). The changes in plasma proteins concentrations are often used for diagnostic purposes (Andonova, 2002; Marshall, 1994). In order to assess the usefulness of fibrinogen as a protein from second group of APPs and as a measure of the acute phase response in rabbits, primary we need to know the reference range of the concentration of fibrinogen in male and female New Zealand white rabbits during different months of their life. Therefore the aim of the study was to measure the alteration in fibrinogen concentration during first three months of their life.

Material and methods

The experimental procedure was approved by the Ethic Committee at the Faculty of Veterinary Medicine. The experiments were carried out on 12 New Zealand white rabbits divided in 2 groups of 6 rabbits in each: I[st] group – 6 male and IInd group – 6 female at ages of 1, 2 and 3 months. All animals were born from healthy doe rabbits and we took blood from them in the rabbitry. They were fed with pelleted feed according to their age and have had free access to the tap water. The concentrations of fibrinogen in male and female rabbits at different ages of this experiment were compared with the concentration of fibrinogen at the first month. In addition, plasma fibrinogen levels were compared between genders at the same ages.

Blood samples from each rabbit were taken from *v. auricularis externa*. The blood was taken into sterile heparinised tubes and were centrifuged immediately (1,500 min^{-1}, for 15 minutes at 4 °C) to obtain plasma.

Plasma fibrinogen was determined immediately by nephelometry method of Podmore with 10% Na_2SO_4 at λ 570 nm (Todorov, 1972).

The statistical analysis of the data was performed using SPSS 16.0 for Windows (SPSS Inc.). Difference of fibronectin between genders was evaluated by Student t-test, while the differences of fibronectin in the groups between the beginning (month 1) and later periods were assessed by paired *t*-test. All data were expressed as mean ± standard deviation (SD) and the differences were considered significant when $P<0.05$.

Results and conclusions

The time course of plasma fibrinogen concentration is presented in Figure 1. In both genders, plasma fibrinogen content at 1 month was comparable. Later on, significant increase of Fib concentrations was found ($P<0.001$). In male rabbits at the months 2 and 3 Fib was more than 1.6-fold and 1.8-fold higher than that at month 1 ($P<0.01$): 1.16±0.44 and 1.25±0.19 g/l, respectively. In female rabbits the changes of fibrinogen followed the similar pattern. At month 2 and 3, the mean plasma Fib concentrations were 1.3- and 1.-5-fold higher than in month 1: 0.90±0.11 and 1.04 ±0.08 g/l, respectively ($P<0.01$).

At the beginning of experiment, at month 1, fibrinogen concentration was equal in both genders, while during the months 2 and 3 fibrinogen concentration in male rabbits was significantly higher as compared to female rabbits ($P<0.01$).

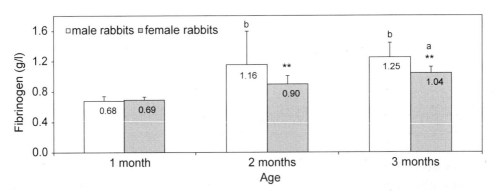

Figure 1. The levels of fibrinogen at different ages in male and female animals. Data are presented as mean±SD ([a]P<0.01, [b]P<0.001 – significance of differences between the initial measurement at month 1 and later periods in both groups; ** P<0.01 – significance of differences between male and female animals in particular periods).

The acute phase response (APR) refers to a nonspecific and complex reaction of an animal that occurs shortly after any tissue injury. The APR includes changes in leukocytes and several plasma proteins as fibrinogen and C-reactive prorein (CRP); these responses are thought to be initiated and regulated systematically. In contrast some facets of the APR, including fever, enhanced sleep, social withdrawal, increases in ceruloplasmin and certain hormonal changes, are mediated by the brain (Shoham *et al.*, 1987). The acute phase proteins are plasma proteins which increase in concentration during the acute phase reaction to infection (Alsemgeest, 1994; Eckersall, 2000).

In animals, several classifications have been proposed on the basis of the human model for post infection APP behavior. Haptoglobin (Hp), ceruloplasmin and fibrinogen were affirmed to be the major positive APPs in rabbits (Petersen *et al.*, 2004). According to cited researchers, Hp increased more than 10 times in rabbits, while fibrinogen – between 2 and 10 times. In our study we present data of plasma fibrinogen concentration in early age of healthy rabbits of both sexes. These data could be used as referent data when studying the changes of this APP in different inflammatory conditions in rabbits.

In conclusion, the present results show that fibrinogen concentration increased with age. In addition, after 1 month of age plasma fibrinogen levels in male rabbits are significantly higher than in female.

References

Alsemgeest, S.P.M., 1994. Blood concentration of acute phase proteins in cattle as markers for disease, Utrecht university, the Nederland, Utrecht.

Andonova, M., 2002. Role of innate defence mechanisms in acute phase response against gram-negative agents. Bulg J Vet Med 5: 77-92.

Eckersall, P.D., 2000. Acute phase proteins as markers of infection and inflammation; monitoring animal health, animal welfare and food safety. Irish Vet.J 53: 307-311.

Kushner, I., Rzewnicki, D. and Samols, D., 2006. What does minor elevation of C-reactive protein signify? Am J Med 119: e17-28.

Marshall, W., 1994. Plasma proteins. In: W. Marshall (Ed.), An illustrated textbook of clinical chemistry, 2nd ed. Mosby, Mosby-Year Bool Europe Limited, London, UK, pp. 210-221.

Petersen, H.H., Nielsen, J.P. and Heegaard, P.M., 2004. Application of acute phase protein measurements in veterinary clinical chemistry. Vet Res 35: 163-187.

Shoham, S., Ahokas, R.A., Blatteis, C.M. and Krueger, J.M., 1987. Effects of muramyl dipeptide on sleep, body temperature and plasma copper after intracerebral ventricular administration. Brain Res 419: 223-228.

Stanley, B. and Van Eyk, J.E., 2005. Clinical proteomics and technologies to define and diagnose heart disease. Walsh. Mech. Cardio 7: 651-665.

Todorov, J., 1972. Nephelimetric determination of fibrinogen (method of Podmore). In: Clinical Laboratory Technics. Sofia, Medizina and Fizkultura, 250.

Part III
Proteomics analysis of food from animal origin

Using shotgun proteomics to understand the effect of feed restriction on the *Ovis aries* wool proteome

André Martinho Almeida[1,2#], Jeffrey E. Plowman[3#], Duane P. Harland[3], Ancy Thomas[3], Tanya Kilminster[4], Tim Scanlon[4], John Milton[5], Johan Greeff[4], Chris Oldham[4] and Stefan Clerens[3]*
[1]*IICT – Instituto de Investigação Científica Tropical and CIISA – Centro Interdisciplinar de Investigação em Sanidade Animal, Lisboa, Portugal; aalmeida@fmv.utl.pt*
[2]*Instituto de Tecnologia Química e Biológica / Universidade Nova de Lisboa and IBET – Instituto de Biologia Experimental e Tecnológica, Oeiras, Portugal*
[3]*AgResearch Ltd, Lincoln, New Zealand*
[4]*Department of Agriculture and Food Western Australia, Perth, WA, Australia*
[5]*University of Western Australia, Crawley, WA, Australia*
[#] *These authors contributed equally to this publication*

Introduction

Sheep wool is an important animal fibre. It is produced throughout the world and assumes a strategic role in the economies of several nations, particularly in the southern hemisphere: Argentina, South Africa and particularly Australia and New Zealand. Wool fibre is obtained solely from sheep (*Ovis aries*), the highest value raw wool coming from pure Merino stock. Numerous factors are known to significantly affect animal fibre production and quality: season, genetics, age, altitude and nutritional factors among many others. Season and nutrition are frequently intertwined factors as seasonal weight loss results from poor quality feed and the availability of pastures in the dry season, and is a major constraint to sheep production in both tropical and Mediterranean regions. Interestingly, and to the best of our knowledge, no studies seemed to be available on the effect of seasonal weight loss or feed restriction on either wool quality traits or the wool proteome itself.

In this study we examined the effects of experimentally induced weight loss on wool protein profiles in Australian Merino ram lambs using isobaric tags for relative and absolute quantitation (iTRAQ). By demonstrating the influence of weight loss on the wool proteome, the results obtained have the potential to provide an insight into the influence of weight loss on wool properties from both a physiological and a commercial point of view.

Material and methods

Animals and experimental conditions have been previously described (Scanlon *et al.*, 2013). All protocols, regarding protein extraction, iTRAQ labelling and analysis used in this work have been thoroughly described previously (Almeida *et al.*, 2014) and are schematized in Figure 1.

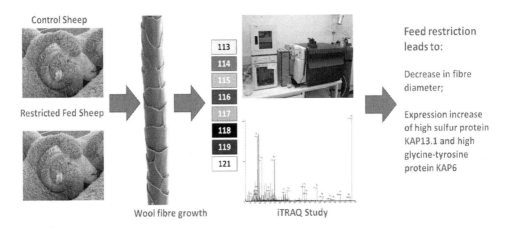

Control Sheep

Restricted Fed Sheep

113
114
115
116
117
118
119
121

Feed restriction leads to:

Decrease in fibre diameter;

Expression increase of high sulfur protein KAP13.1 and high glycine-tyrosine protein KAP6

Wool fibre growth   iTRAQ Study

Figure 1. Schematization of the strategy used to study the effect of weight loss on the wool proteome.

Results and discussion

Live weight changes for the two experimental groups have been previously published Scanlon *et al.* (2013). Briefly, animals in the R group lost 15% of their initial live weight, whereas for animals in the C group there was a 12% increase in live weight. In the case of the wools, the fibre diameter in the animals fed the restricted diet were found to be significantly lower than those fed the growth diet. Regarding the iTRAQ analysis, a total of 11 trichocyte (hair) keratins, 11 keratin associated proteins (KAPs), two epithelial keratins, and two other non-keratinous proteins were identified and quantitated using this approach. Based on the results calculated using all of the peptides, most of the control:restricted diet ratios for the trichocyte keratins were greater than unity, though only three of them were significant, those for K32, K35 and K86 but none were considered to be significantly different when only unique peptides were taken into consideration. In contrast, for a number of KAPs the ratios were found to be significantly below unity. This included the HSP KAP13.1 based on five peptides, when either all or only unique peptides were taken into account. Three high glycine-tyrosine proteins (HGTPs) KAP6, KAP6.3 var1 and KAP6.2 were also less than unity based on three peptides, as was the HGTP KAP8.1, though this based on only one peptide from the C-terminus which was labelled at either one or two sites. For all these proteins this suggests that there was a higher abundance of these proteins in the restricted diet animals.

The combination of the results herein presented with a decrease in fiber diameter shown by restricted-feed animals (data not shown) seems to indicate that wool from animals subjected to SWL and hence may have a reduced prickle but that the increase in HGTPs in restricted-feed animals may drive a decrease in crimp and hence a reduction in wearability and appearance retention, thus with strong implications for the wool and wool-derived textile industry (Almeida *et al.*, 2014).

Acknowledgements

Author AM Almeida acknowledges financial support from *Fundação para a Ciência e a Tecnologia* (Lisbon, Portugal) in the form of Grant SFRH/BPD/90916/2012. Animal work was supported by the Department of Agriculture and Food of the Government of Western Australia (Perth, WA, Australia). We are also grateful to Dr David Scobie and Joy Woods, both from AgResearch, for helpful advice and discussions during the preparation of the manuscript. Funding for the New Zealand part of this project was provided by the Ministry of Business, Innovation and Employment (C10X0710) and AgResearch Core Funding (A19115). This work was made possible through COST-STSM-RA – New Zealand-06429 Reciprocal Short Term Scientific Mission (RSTSM) funded by the European Science Foundation through COST (Cooperation through Science and Technology). Authors AM Almeida and JE Plowman are members of COST action FA1002 – Proteomics in Farm Animals (www.cost-faproteomics.org) from which networking funding is acknowledged.

References

Almeida, A.M., Plowman, J.E., Harland, D.P., Thomas, A., Kilminster, T., Scanlon, T., Milton, J., Greeff, J., Oldham, C. and Clerens, S., 2014. Influence of feed restriction on the wool proteome: a combined iTRAQ and fibre structural study. Journal of Proteomics 30, 170-177.

Scanlon, T.T., Almeida, A.M., van Burgel, A., Kilminster, T., Greeff, J.C. and Oldham, C., 2013. Live weight parameters and feed intake in Dorper, Damara and Australian Merino lambs exposed to restricted feeding. Small Ruminant Research 109, 101-106.

Comparative proteomic analysis of muscle tissue from pre-term and term calves

Paula Friedrichs[1], Hassan Sadri[1], Julia Steinhoff-Wagner[2], Harald Hammon[2], Allan Stensballe[3], Emøke Bendixen[4] and Helga Sauerwein[1]*

[1]*Institute of Animal Science, Physiology and Hygiene Unit, University of Bonn, Katzenburgweg 7-9, 53115 Bonn, Germany; sauerwein@uni-bonn.de*
[2]*Leibniz Institute for Farm Animal Biology (FBN), Institute of Nutritional Physiology, Wilhelm-Stahl-Allee 2, 18196 Dummerstorf, Germany*
[3]*Department of Health Science and Technology, University of Aalborg, Fredrik Bajers Vej 3, 9220 Aalborg, Denmark*
[4]*Institute for Molecular Biology, University of Aarhus, Gustav Wieds Vej 10, 8000 Aarhus C, Denmark*

Objectives

Preterm birth implies a high risk of adverse long term outcomes in terms of performance and health for survivors, yet the underlying molecular mechanisms are unclear. Proteomics may help to get a more holistic picture of the reactions in support of the adaptation from intra to extra-uterine life in both normal term (T) and preterm (PT) born neonates. Skeletal muscle is one of the largest organs; it plays a major role in protein and amino acid metabolism. Focusing on dairy calves, we aimed to get first insights into the proteome of muscle tissue from neonatal calves comparing (PT) and (T) calves to identify proteins that are affected by relative maturity at birth.

Materials and methods

Six German Holstein calves either delivered preterm by caesarian section nine days before anticipated calving date (PT, n=3) or were born spontaneously at term (T, n=3), were kept in individual boxes with straw bedding and free access to water. One PT calf was female the other calves within this trial were male. They did not receive colostrum, milk or formula during the first 24 h post natum, but were fed 2 h before slaughter with pooled colostrum at 5% of body weight. After 26 h of life, the calves from both groups were slaughtered. Samples from *M. longissimus dorsi* were dissected and were snap-frozen in liquid nitrogen and stored at -80 °C until analysed. More details of this study were reported previously (Steinhoff-Wagner *et al.*, 2011). Using each 200 mg, the samples were homogenised in 1 ml TES buffer in a TissueLyser (QIAGEN, Venlo, Netherlands) for 3 times for 30 sec at 30 Hz and were then centrifuged at 3,000×*g* for 30 min at 4 °C. The supernatant was mixed with the 6-fold volume of acetone for protein precipitation. After centrifugation for 10 min at 4 °C and 15,000×*g*, the supernatant was discarded and the tubes were stored with open lids for 30 min under an air steam to allow the pellets to dry. Two sets of 4-plexed iTRAQ experiments were designed in order to analyse and compare the proteome of the three samples from both groups and each set was analysed in two technical replicates. Consistently, the iTRAQ reagent 114 was used to label a pool of the muscle tissue samples consisting of an equal protein amount from each sample. The iTRAQ reagents 115, 116 and 117 were used to label the individual tissue samples from the PT and the T group. The samples from both groups were spread over the different sets to avoid set dependent differences. Peptide mixtures for shotgun analyses were generated from the digestion of 50 µg of protein aliquots. The disulphide bonds of the proteins were reduced with 2.5 mM tris-(2-carboxyethyl) phosphine at 60 °C for 1 h, afterwards the cysteine groups were blocked with 10 mM methyl methanethiosulfonate in isopropanol for 10 min. The proteins were digested with trypsin at 37 °C overnight and protein digests were labeled with the iTRAQ reagents (114, 115, 116 and 117) at room temperature for 1 h. Finally, samples were combined to one mixture to create 4-plexed samples. The iTRAQ labeled peptide 4-plexed samples were pre-separated on Agilent 1100 Series capillary HPLC (Agilent, Santa Clara, CA, USA) equipped with a Zorbax Bio-SCX Series II (Agilent), and peptides were eluted with a gradient of increasing NaCl concentration. The resulting peptides were then analyzed on a nanoflow UPLC (Dionex Ultimate3000/RSLC, ThermoFisher Scientific, Waltham, MA, USA) system coupled online by a nanospray ion source (Proxeon, ThermoFisher Scientific) to an Orbitrap Q-Exactive mass spectrometer (ThermoFisher Scientific) as described earlier (Dueholm *et al.*, 2013). Separation of the peptides was achieved on two successive reverse phase columns (Acclaim PepMap100 C18 Nano-Trap, and Column Acclaim PepMap300 C18, ThermoFisher Scientific). The collected MS files were available in the mascot generic format (MGF) and they were analysed using the software ProteinPilot™ (AB SCIEX, Framingham, MA, USA) with the MGF parameters. The generated peak lists were used to interrogate a bovine genome reference set database downloaded from UniProt (http://www.uniprot.org, taxon identifier 9913). As quantification method, the iTRAQ 4-plex was selected. The proteins that were not annotated were blasted by hand using their specific accession ID and the Ensembl project genome database (http://www.ensembl.org). The proteins were classified with the PANTHER classification system (http://www.pantherdb.org) based on their accession ID as described before (Mi *et al.*, 2013). The differences ($P<0.05$) of the fold-changes compared to the pool sample of the identified proteins between the PT and T group were tested with student's t-test in Excel 2007 (Microsoft, Redmond, WA, USA).

Results and discussion

In total 370 proteins were identified and quantified in muscle tissue of T and PT calves, whereby this value includes only proteins that where present in both sets performed for muscle tissue. In accordance with other proteomic studies on muscle tissue of cattle, we also observe large portions of creatine kinase M-type, pyruvate kinase, phosphoglucomutase-1, myosin-1, and beta-enolase relative to all other detected proteins (Cooper-Prado *et al.*, 2014; Kuhla *et al.*, 2011). Molecular functional analysis using PANTHER identified the most relevant biological functions for 321 proteins of the total protein dataset. The most relevant molecular functions were observed for catalytic activity and binding (45.6% and 26.6% of the proteins are involved in the respective function). As shown in Table 1, in muscle tissue the abundance of 16 proteins differed between the T and PT calves, of which 11 proteins were higher in the calves from the T group.

Table 1. List of proteins, that differed in the pre-term (PT) vs term (T) calves. The shown values are means of the fold-changes compared to the pool sample labeled with the iTRAQ reagent 114.

Acc. ID[1]	Protein name	PT	T	*P*-value
tr|F1N757	Titin[2]	0.4587	1.0670	0.0481
sp|Q3ZCH0	Stress-70 protein, mitochondrial	0.2290	1.3391	0.0007
sp|P00760	Cationic trypsin	0.1253	0.6329	0.0007
tr|G3MZU6	IGFN1[2,3]	0.5015	2.1744	0.0307
tr|A5D7Q4	CSDA protein	0.3705	1.7324	0.0369
tr|F1N690	Dihydrolipoyllysine acetyltransferase PDC[4]	0.5391	1.1638	0.0150
sp|O02691	3-hydroxyacyl-CoA DHG[5] type-2	1.2662	4.3791	0.0459
sp|Q2T9S4	Phosphoglycolate phosphatase	0.1650	1.1388	0.0051
sp|Q2TBQ3	Guanidinoacetate N-methyltransferase	1.0803	0.9288	0.0483
sp|Q148N0	2-oxoglutarate DHG[5], mitochondrial	0.4160	1.1306	0.0320
sp|P81947	Tubulin alpha-1B chain	1.0745	0.8240	0.0120
tr|F6QLM5	Glycogenin 1 (GYG1)[2]	0.2631	1.2441	0.0201
sp|P08814	Parathymosin	0.1728	1.1176	0.0189
sp|Q32KW2	UBX domain-containing protein 1	1.4344	0.8620	0.0436
tr|Q0VCY8	Phosphoprotein enriched in astrocytes 15	1.2767	0.8291	0.0311
sp|P25417	Cystatin-B	1.8846	1.0598	0.0040

[1] Accession ID.

[2] Blasted and annotated by hand based on the specific accession ID.

[3] Immunoglobulin-like and fibronectin type III domain containing 1.

[4] Pyruvate dehydrogenase complex.

[5] Dehydrogenase.

A classification by molecular function of all the 16 regulated proteins identified six proteins with catalytic activity. Except cystatin B, a protease inhibitor, and cationic trypsin, a serine protease, all regulated catalytic active proteins were higher in the T calves suggesting that there is a higher rate of chemical reactions in T vs PT calves. It was also notable by pathway classification that three regulated proteins are involved in the gonadotropin releasing hormone receptor pathway, whereat we observed no differences in the PT group between the female animal and the male ones. This pathway is modulated in response to gonadotropin releasing hormone, sex steroids and gonadal peptides, as well as across sexual maturation (Harrison *et al.*, 2004). The abundance of titin, the largest protein found in muscle (Huff-Lonergan *et al.*, 1995) was greater in T than in PT calves' muscle.

Conclusion

The results from the investigation of tissue from *M. longissimus dorsi* indicate that there is a difference in the proteome of PT and T calves, in particular regarding catalytic activity. The comparative analysis of muscle tissue form PT and T calves also points to a difference in the gonadotropin releasing hormone receptor pathway and warrants further investigation.

References

Cooper-Prado, M.J., Long, N.M., Davis, M.P., Wright, E.C., Madden, R.D., Dilwith, J.W., Bailey, C.L., Spicer, L.J. and Wettemann, R.P., 2014. Maintenance energy requirements of beef cows and relationship with cow and calf performance, metabolic hormones, and functional proteins. Journal of Animal Science 92(8):3300-15.

Dueholm, M.S., Sondergaard, M.T., Nilsson, M., Christiansen, G., Stensballe, A., Overgaard, M.T., Givskov, M., Tolker-Nielsen, T., Otzen, D.E. and Nielsen, P.H., 2013. Expression of Fap amyloids in Pseudomonas aeruginosa, P. fluorescens, and P. putida results in aggregation and increased biofilm formation. MicrobiologyOpen 2(3): 365-382.

Harrison, G.S., Wierman, M.E., Nett, T.M. and Glode, L.M., 2004. Gonadotropin-releasing hormone and its receptor in normal and malignant cells. Endocrine-Related Cancer 11(4): 725-748.

Huff-Lonergan, E., Parrish, F.C., Jr. and Robson, R.M., 1995. Effects of postmortem aging time, animal age, and sex on degradation of titin and nebulin in bovine longissimus muscle. Journal of Animal Science 73(4): 1064-1073.

Kuhla, B., Nürnberg, G., Albrecht, D., Gors, S., Hammon, H.M. and Metges, C.C., 2011. Involvement of skeletal muscle protein, glycogen, and fat metabolism in the adaptation on early lactation of dairy cows. Journal of Proteome Research 10(9): 4252-4262.

Mi, H., Muruganujan, A., Casagrande, J.T. and Thomas, P.D., 2013. Large-scale gene function analysis with the PANTHER classification system. Nature Protocols 8(8): 1551-1566.

Steinhoff-Wagner, J., Gors, S., Junghans, P., Bruckmaier, R.M., Kanitz, E., Metges, C.C. and Hammon, H.M., 2011. Maturation of endogenous glucose production in preterm and term calves. Journal of Dairy Science 94(10): 5111-5123.

High frequencies of the α_{S1}-casein zero variant and its relation to coagulation properties in milk from Swedish dairy goats

Monika Johansson[1]*, Madeleine Högberg[2] and Anders Andrén[1]

[1]Department of Food Science, Uppsala BioCenter, Swedish University of Agricultural Sciences, Sweden, 750 07 Uppsala, Sweden; monika.johansson@slu.se

[2]Department of Anatomy, Physiology and Biochemistry, Swedish University of Agricultural Sciences, Uppsala, Sweden

Introduction

Milk, which contains a higher proportion of α_{S1}- and κ-CN, positively affects the cheese making properties, for example by having a higher total protein, fat and calcium content (Clark and Sherbon, 2000). The Norwegian Landrace goats have been shown to belong to a population in which some dairy goats through a mutation lost the ability to produce α_{S1}-CN (Hayes *et al.* 2006). The frequency has been measured in several studies to over 70% of this 'zero'-variant, which significantly reduces the milk casein and fat content and thereby the cheese yield (Devold *et al.* 2011).

As many as 18 different variants of α_{S1}-CN have been recognized in goat breeds (Caroli *et al.* 2007). The ability to produce α_{S1}-CN is mainly controlled by the alleles number 1, 3 and 6, of which the A-variant (allele 6) produces 3.6 g of α_{S1}-CN per litre, the G-variant (allele 3) 0.6 g/l/allele and the D-variant (allele 1) not any α_{S1}-CN at all (Martin *et al.* 1999). Depending on the ability to produce α_{S1}-CN the goats use to be classified into strong-, medium-, weak- or zero-goats, where the strong producing goats are the most suitable for milk production for cheese making with good rennet coagulation properties and high cheese yield. In most goat breeds of Southern Europe, the ratio of weak and zero variants are very low. Exceptions are the Spanish Canaria and the Italian Garganica breeds, which have the gene frequencies 20% and 23%, respectively, of the zero variant (Caroli *et al.* 2007).

Thus, the ability of goats to produce milk with high contents of α_{S1}-CN has huge economical impact for dairy goat farms producing goat cheeses. Since the Swedish Landrace goats are closely related to the Norwegian Landrace and bucks from Norway have been used in the breeding of Swedish goats, it is very likely that even the Swedish goats carry the gene for zero-synthesis of α_{S1}-CN. The objective of this study was therefore to provide a random survey of the Swedish goat population with the aim to identify the frequency of Swedish goats producing low levels of α_{S1}-CN and the determination of the relation between α_{S1}-CN and coagulation properties.

Material and methods

Content of a_{S1}-CN

Milk samples were collected from 283 goats from 28 farms from ten different geographical regions of Sweden. Before the analyses, the samples were defatted at 3,000×g at 4 °C for 10 min. Electromigration of the goat milk proteins was carried out with a CZE (G-1600AX, Agilent Technologies Co., SE-164 94, Kista, Sweden), controlled by Chemstation software version A 10.02. Separations were performed according to the methods described by Åkerstedt *et al.* (2014). The calculation of relative concentrations of the individual proteins was based on the peak area and expressed as a percentage of the total areas recorded for all peaks in the electropherogram. The percentage of α_{S1}-CN compared to the other caseins were calculated and then classified as strong (15-25%), medium (7-14.9%) and low (0-6.9%) to get a grouping on the expression of α_{S1}-CN.

Rheological analyses

62 goats were further investigated for the milk coagulation. All samples were defatted at 3,000×g at 4 °C for 10 min before evaluations. Elastic modulus G´ and viscous modulus G´´ of the milk samples were continuously measured using Bohlin CVOR-150-900 rheometer (Malverin Instruments Nordic AB, Uppsala, Sweden). The rheometer was equipped with a cup and a concentric cylinder 25 and 28 mm in diameter respectively at a high of 40 mm. Temperature was controlled by a peltier element. Aliquots of milk (12 ml) were equilibrated to 35 °C in a water bath for 10 min. Chy-max Ultra (Chr. Hansen A/S, Hørsholm, Danmark) was added at a concentration of 75 IMCU/ml. Gel forming was followed for 30 min with an oscillation frequency of 1 rad/s and a strain of 0.01, which was well within the linear viscoelastic region of milk gels. Measurement frequency was set to 8 s. Coagulation time (Ct) was measured from the point of the enzyme addition until reaching the 1 Pa value. Gel firmness (G´) of the developing gel was plotted against time after 20 minutes (G_{20}).

Results and discussion

Distribution of a_{S1}-CN in the goat population

There was a very high frequency of the low expression of α_{S1}-CN in Swedish dairy goats. Of 283 analysed goat milk samples, 185 (65%) had an expression of α_{S1}-CN below 7%, which means that these goats produce very little α_{S1}-CN, if any. Only 35 goats (12%) produced high levels of α_{S1}-CN and 63 (22%) were considered medium producers of α_{S1}-CN. Protein profiles of goat milk with high and low expression of α_{S1}-CN based on capillary zone electrophoresis (CZE) electropherograms are presented in Figure 1A and 1B, respectively.

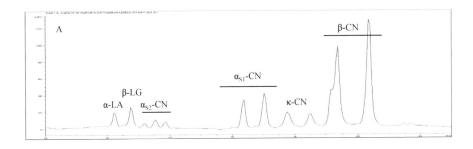

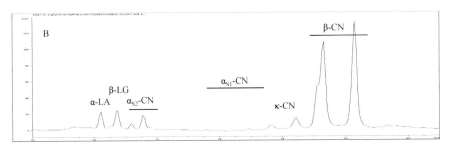

Figure 1. Representative electropherograms of goat milk protein profile. (A) Protein profile with a high expression of α_{S1}-CN. (B) Protein profile with a low expression of α_{S1}-CN. The individual proteins are indicated. α-LA, α-lactalbumin; β-LG, β-lactoglobulin; α_{S1} CN, α_{S1}-casein; α_{S2}-CN, α_{S2}-casein; κ-CN, κ-casein; β-CN, β-casein.

Coagulation properties

Distinct differences for both the coagulation time and gel firmness were observed. We could show, that the coagulation time for the low group is 15% longer than the group with high levels of α_{S1}-CN (Table 1). Milk with low α_{S1}-CN content showed 60% lower gel firmness ($P<0.01$) compare to the milk with high α_{S1}-CN (Table 1).

Table 1. Coagulation properties and pH of milk from Swedish Landrace goats. Average coagulation time (Ct), G_{20}, gel firmness after 20 minutes in milk with low (0-6.9%), medium (7-14.9%) and high (15-25%) α_{S1}-casein (α_{S1}-CN) content. Values are mean ± standard error.

	Content of α_{S1}-CN			*P*-value
	Low	**Medium**	**High**	
Ct (s)	905±5	780±6	767±7	ns
G_{20} (Pa)	18±4	28±4	44±6	**

**P<0.01; ns = non-significant.

Conclusion

The frequency of Swedish goats producing low levels of α_{S1}-CN was high and the expected relationship between Swedish and Norwegian dairy goats is thus proved in the ability to produce this protein. From the findings of this study it can be concluded that the coagulation properties of goat milk is highly depended on α_{S1}-CN content. If the expression of this protein is high, it is accompanied by shorter coagulation time and better gel firmness. Therefore, goat milk with high level of α_{S1}-CN demonstrates appropriate technological properties and is better suited for cheese production purposes than milk with low α_{S1}-CN content.

References

Åkerstedt, M., Wredle, E., Lam, V., Johansson, M., 2012. Protein degradation in bovine milk caused by *Streptococcus agalactiae*. J. Dairy Res. 79, 297-303.

Caroli, A., Chiatti, F., Chessa, S., Rignanese, D., Ibeagha-Awemu, E.M., Erhardt. G., 2007. Characterization of the Casein gene complex in West African goats and description of a new αS1-Casein polymorphism. J. Dairy Sci. 90, 2989-2996.

Clark, S., Sherbon. J.W., 2000. Alpha(s1)-casein, milk composition and coagulation properties of goat milk. Small Ruminant Res. 38, 123-134.

Devold, T.G., Nordbø, R., Langsrud, T., Svenning, C., Brovold, M.J., Sørensen, E.S., Christensen, B., Ådnøy, T., Vegarud. G.E., 2011. Extreme frequencies of the αs1-casein 'null' variant in milk from Norwegian Dairy Goats – implications for milk composition, micellar size and renneting properties. Dairy Sci. Technol. 91, 39-51.

Hayes, B., Hagesæther, N., Ådnøy, T., Pellerud, G., Berg, P.R., Lien, S., 2006. Effects on production traits of haplotypes among casein genes in Norwegian goats and evidence for a site of preferential recombination. Genetics 174, 455-464.

Martin, P., Ollivier-Bousquet, M., Grosclaude, F., 1999. Genetic polymorphism of caseins: a tool to investigate casein micelle organization. Int. Dairy J. 9, 163-171.

First characterization of the goat mammary gland proteome secretory tissue using shotgun proteomics

Joana R. Lérias[1*], Lorenzo E. Hernández-Castellano[2], Noemí Castro[3], Anastasio Argüello[3], Juan Capote[4], Alan Stensballe[5], Jeffrey E. Plowman[6], Stefan Clerens[6], Emoke Bendixen[7] and André M. de Almeida[1,8]

[1]Instituto de Investigação Científica Tropical, Centro de Veterinária e Zootecnia, Faculdade de Medicina Veterinária, Av. Univ. Técnica, 1300-477 Lisboa, Portugal and IBET, Av. República, 2780-157 Oeiras, Portugal; jrlerias@fc.ul.pt

[2]Vetsuisse Faculty, Veterinary Physiology, University of Bern, Bern, Switzerland

[3]Facultad de Veterinaria, Universidad de Las Palmas de Gran Canaria, Arucas, Spain

[4]ICIA – Instituto Canario de Investigaciones Agrarias, Valle Guerra, Tenerife, Spain

[5]Aalborg University, Aalborg, Denmark

[6]AgResearch Ltd, Lincoln, Canterbury, New Zealand

[7]University of Aarhus, Aarhus, Denmark

[8]ITQB – Instituto de Tecnologia Química e Biológica, Oeiras, Portugal and CIISA – Centro Interdisciplinar de Investigação em Sanidade Animal, Lisboa, Portugal

Introduction

The animal industry is gaining importance in developing countries, with a special emphasis on goats. These animals are considered to be resilient to several adversities common in these countries, such as poor feed availability during the dry season. For this reason they are considered an interesting alternative for the supply of dairy products for human consumption. In addition, dairy production is considered to be an essential tool to overcome social and economic issues in developing countries (McDermott *et al.*, 2010). It is therefore crucial to understand goat mammary gland physiological and anatomical functions. Specifically, characterization of the mammary gland proteome helps to understand differences between animals under several conditions, e.g. under different nutrition levels or different milking frequencies. At a structural level, this organ suffers modifications along lactation and also according to different milking frequencies and lactation number (Lérias *et al.*, 2014). However, to our knowledge there are no descriptions of differences at the proteome level, enhancing the importance of such results.

Material and methods

Sample collection

Mammary gland samples were obtained from a study conducted at the experimental farm of the Faculty of Veterinary Medicine of the ULPGC – University of Las Palmas de Gran Canaria (Gran Canaria, Spain) as previously described by Lérias *et al.* (2013). At the end of the trial, mammary gland biopsies were obtained under competent veterinary supervision from all the animals and then were labeled and frozen at -80 °C until further analysis.

Sample preparation

First, 200 mg of each mammary gland sample were homogenized with 1 ml of TES buffer using an ULTRA-TURRAX at 12.000 rpm. After homogenization, samples were centrifuged at 10,000×g for 30 min at 4 °C. Protein concentration of the supernatant was determined with the Quick Start™ Bradford Protein Assay (Bio-Rad, Hercules, CA, USA), using BSA as standard reference. After quantitation, 100 µg of protein from each sample were obtained after precipitation with 6 volumes of ice-cold acetone (-20 °C) at 15,000×g for 10 min under refrigeration conditions (4 °C).

iTRAQ design, protein digestion and iTRAQ labeling

iTRAQ 4-plex reagent 114 was used to label a pooled mammary gland sample consisting of an equal protein amount of each of the 17 mammary gland samples. The other iTRAQ reagents (115, 116 and 117) were used to label the individual mammary gland samples from *Majorera* (5 animals) and *Palmera* (4 animals) breeds. Cysteine residues were reduced with TCEP-IICL at 60 °C for 1 h and then blocked with 10 mM MMTS at room temperature for 10 min. Proteins were then digested with trypsin (1:10 w/w) at 37 °C overnight. Digested proteins were labeled with the iTRAQ reagents (114, 115, 116 and 117) and incubated at room temperature for 1 h. Finally, the samples were mixed in a single tube and mixed by vortexing.

SCX fractionation

Peptide mixtures generated from the digestion of 50 µg of protein were injected into an Agilent 1100 Series capillary HPLC equipped with a Zorbax Bio-SCX Series II, and peptides were eluted with a gradient of increasing NaCl solution. Fractions were collected every minute for 65 minutes and then were combined according to their peptide loads into 9-10 pooled samples. The pooled SCX fractions were further separated by a reverse phase liquid chromatography using an Agilent 1100 Series nano-flow HPLC system.

LC-MS/MS

The peptides were analyzed on a nanoflow UPLC (Dionex Ultimate3000/RSLC, ThermoFisher Scientific, Waltham, MA) system coupled online by a nanospray ion source (Proxeon, ThermoFisher Scientific) to an Orbitrap Q-Exactive mass spectrometer (ThermoFisher Scientific). The peptides were separated on two successive reverse phase columns (Acclaim PepMap100 C18 Nano-Trap, and Column Acclaim PepMap300 C18, ThermoFisher Scientific) using a linear gradient (10-35% acetonitrile in 35 min) and a constant flow rate of 300 nl/min. The mass spectrometer was operated in a data-dependent mode to switch between full MS scans and tandem MS/MS. Fragmentation was performed using high-energy collision induced dissociation and sequenced precursor ions were dynamically excluded for 30 seconds. The raw mass spectrometry files were analyzed and exported as mgf files using Thermo Proteome Discover (version 1.3.0.339).

Protein identification

The collected MS files were analyzed in the Mascot software to interrogate an in-house assembled goat database consisting of sequences from NCBInr (22/05/2014, 32464 sequences). Search parameters were as follows: Enzyme: Trypsin; Fixed modifications: Methylthio (C); Variable modifications: iTRAQ4plex (K), iTRAQ4plex (N-term), iTRAQ4plex (Y), Deamidated (NQ), Oxidation (M), Delta:H(2)C(2) (N-term); Peptide tolerance: 10 ppm; Fragment tolerance: 0.4 Da; Missed cleavages: 1; Instrument: ESI-TRAP.

Only peptides with an ions score above 20 and proteins with a score above 80 were considered. The cellular localization as well as the metabolic pathways in which each identified protein is involved were ascertained by searching the obtained GI numbers in the UniProt website.

Results and discussion

The pathways/components and cellular localization of the mammary gland proteome are presented in Figures 1 and 2, respectively. A total of 774 proteins were detected among all animals, most of them being important for protein metabolism (35%) and located in the cytoplasm (60%). Additionally, we have also identified several proteins linked to immune and stress response (4% and 3%, respectively), which is expected considering that the mammary gland organ has a high muscular component (explains the high content of proteins important for protein metabolism) and a high prevalence of bacterial infection (proteins linked to immune and stress response). Additionally, these results also showed the presence of lipid and carbohydrate metabolism-related

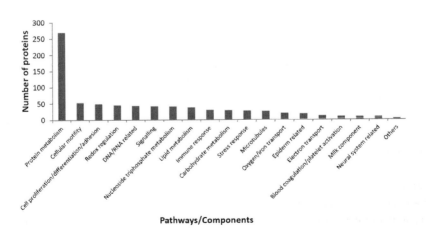

Figure 1. Pathways/components of mammary gland proteome.

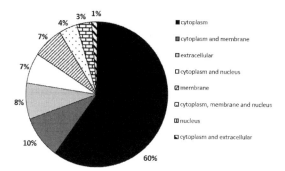

Figure 2. Cellular localization of mammary gland proteins.

proteins, as well as signaling and nucleoside triphosphate metabolism. Finally, there was also the detection of several epidermal related proteins, which are likely due to contamination during the biopsy excision process.

Future perspectives

The results herein presented are the first characterization of the goat mammary gland proteome. As previously described, small ruminants' mammary gland suffers several changes during lactation at a structural point of view and, possibly, also at a proteomic point of view. It would be of interest to study the evolution of not only the mammary gland proteome, but also the protein expression profiles along the length of the lactation cycle, in order to improve our knowledge in this process. Indeed, and as the mammary gland of these animals is highly susceptible to bacterial infection and inflammation (mastitis), it would be important to understand if throughout the lactation cycle there are important differences regarding the expression of proteins involved in defense mechanism, possibly allowing the development of new strategies for mastitis prevention.

Acknowledgements

This project was funded by the research project PTDC/CVT/116499/2010 (FCT, Lisbon, Portugal) and by the Cost Action FA 1002 through a Short-term Scientific Mission under the original project name of 'Goat mammary gland proteomic study in two breeds from the Canary Islands with different adaptations to SWL' (reference COST-STSM-ECOST-STSM-FA1002-211013-032143).

References

Lérias, J.R., Hernández-Castellano, L.E., Morales-delaNuez, A., Araújo, S.S., Castro, N., Argüello, A., Capote, J. and Almeida, A.M., 2013. Body live weight and milk production parameters in the Majorera and Palmera goat breeds from the Canary Islands: influence of weight loss. Tropical Animal Health and Production 45, 1731-1736.

Lérias, J.R., Hernández-Castellano, L.E., Suárez-Trujillo, A., Castro, N., Pourlis, A. and Almeida, A.M., 2014. The mammary gland in small ruminants: major morphological and functional events underlying milk production – a review. Journal of Dairy Research 81, 304-318.

McDermott, J.J., Staal, S.J., Freeman, H.A., Herrero, M. and Van De Steeg J.A., 2010. Sustaining intensification of smallholder livestock systems in the tropics. Livestock Science 130, 95-109.

Identification of potential biomarkers of animal stress in the muscle tissue of pigs caused by different animal mixing strategies

A. Rubio-González[1], M. Oliván[2*], Y. Potes[1], D. Illán-Rodríguez[1], I. Vega-Naredo[1], V. Sierra[2], B. Caballero[1], E. Fàbrega[3], A. Velarde[3], A. Dalmau[3], F. Díaz.[2] and A. Coto-Montes[1]

[1]Department of Morphology and Cellular Biology, Faculty of Medicine, University of Oviedo, Julián Clavería s/n, 33006 Oviedo, Principado de Asturias, Spain
[2]Servicio Regional de Investigación y Desarrollo Agroalimentario (SERIDA), Apdo. 13, 33300 Villaviciosa, Principado de Asturias, Spain; mcolivan@serida.org
[3]Institut de Recerca i Tecnologia Agroalimentàries (IRTA), Veïnat de Sies, s/n, 17121 Monells, Girona, Spain

Understanding the mechanisms that influence the post-mortem conversion of muscle into meat could lead to the identification of biomarkers of animal welfare and meat quality. Mixing unfamiliar animals may be an important ante- and peri-slaughter stressor. The objective of this work was to study the post-mortem evolution of potential biomarkers of autophagy and oxidative stress in the *Longissimus dorsi* muscle of male ((Large White × Landrace) × Duroc) pigs subjected to different management treatments that promote psychological stress, as mixing unfamiliar animals at farm and/or during transport and lairage prior to slaughter. We demonstrated that mixing animals produces increased muscle oxidative stress and triggers autophagy, as a mechanism for survival of the muscle cells. This affected to the post-mortem muscle metabolism, with significant changes in some metabolic proteins like L-lactate dehydrogenase and adenylate kinase isoenzyme 1, with important effects on meat quality traits. From these results, we propose that monitoring the post-mortem evolution in the muscle of main biomarkers of the cell antioxidant defence (Total Antioxidant Activity, Superoxide Dismutase Activity, Catalase Activity) and autophagy (Beclin 1, LC3II/LC3I ratio) could serve as indicators of animal stress and meat quality.

Novelty and tradition: when proteomics meets Nero di Parma ham

Gianluca Paredi[1*], Samanta Raboni[1], Roberta Virgili[2], Alberto Sabbioni[3] and Andrea Mozzarelli[1,4]

[1]Department of Pharmacy, Interdepartmental Center Siteia Parma, University of Parma, Parma, Italy; gianluca.paredi@unipr.it

[2]Stazione Sperimentale per l'Industria delle Conserve Alimentari, Parma, Italy

[3]Department of Veterinary Science, University of Parma, Italy

[4]National Institute of Biostructures and Biosystems, Rome, Italy

Introduction

Proteomics approaches have been successfully applied to characterize at the protein level the modifications that accompany the transformation of raw pork meat in dry-cured ham in the technological process (Paredi *et al.*, 2012). *Nero di Parma* is a black coated pig population typical of Parma that was reintroduced in the 90s and to date accounts up to 8,500 total and 900 living individuals (Sabbioni *et al.*, 2010). Among all the meat products of *Nero di Parma*, dry cured ham is the most famous and is manufactured according to established technological steps mainly based on empirical observations and traditional recipes. Meat salt intake, dehydration and water activity decrease are some of the key events that lead to a food product that can be stored at room temperature for several months with typical sensory and quality traits. In previous studies we successfully applied proteomics approaches to characterize the modification that pig meat undergoes during the production of cooked and dry cured ham (Paredi *et al.*, 2013; Pioselli *et al.*, 2011). The aim of the present study was the proteomic characterization of exudates generated from *Nero di Parma* meat during the salting phase.

Materials and methods

Four fresh pork meat legs of *Nero di Parma* were selected in a local slaughterhouse. Legs were salted with the classic two steps process, using a mixture of wet and dry salt (NaCl). Hams were manually salted using two kinds of salt (a mixture of 2- and 3-mm grain size): wet salt (nearly 15% added water) was rubbed on ham rind, while dry salt was used for the unskinned ham part. 5% salt (calculated on ham weight) was used for the first salting, while an addition of 2.8% salt was made in the second salt. Salted hams were placed in a room operating at controlled relative humidity of 80-90% at 1-3 °C. At the end of the first salting phase (6 days) hams were washed from salt, rubbed with new salt, and stored under the same environmental conditions of the first salting for 12 days. The whole salting treatment lasted 18 days. The exudate dripped out from hams during the whole process was collected and were stored at -80 °C before processing. Subsequently, the exudates were centrifuged in order to remove insoluble components. A desalting step was carried out dialyzing the samples against 40 mM Tris 0.5% SDS pH 7.4 for 24 hours. Finally, in order to concentrate the protein extracts, 200 µl aliquots were treated by quantitative acetone precipitation overnight. Protein concentration was determined with the Bradford method. The exudates were analyzed using both 1D-PAGE and 2D-PAGE gels. The 2D gels were compared using PDQUEST™ (BIORAD) software.

Results and discussion

A first comparison of the proteome of exudates at day 1, 6 and 18 was carried out with 1-DE (Figure 1), showing quantitative difference in several bands.

Specifically, 5 bands show an increasing intensity from day 1 to day 18, whereas 4 bands show an opposite behavior. Finally, a single band at low molecular weight appeared at day 18 day. This band might be a potential marker of the proteolysis that takes place during the salting phase. The comparison with our previous investigation on pig meat indicates that the 1D-PAGE profiles from *Nero di Parma* samples show a higher variability in band intensities as a function of time. In order to obtain a better understanding of modifications triggered by salt, a 2D-PAGE analysis of the proteins in exudates collected at day 1, 6 and 18 was carried out in the pH range 4-7 (Figure 2). An average of 219, 227 and 251 spots were identified, respectively. 126 spots were common to all three samples and 23 were differentially extracted. Particularly, 13 spots were more concentrated

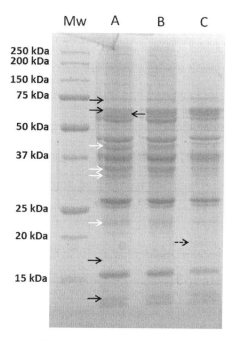

Figure 1. SDS-PAGE of exudates collected at 1 (A), 6 (B) and 18 (C) days. The bands more intense in 1 day sample are marked with white arrows. The black arrows mark protein bands more intense in the 18 days sample. Finally, the black dotted arrow mark the band identified only in 18 days sample.

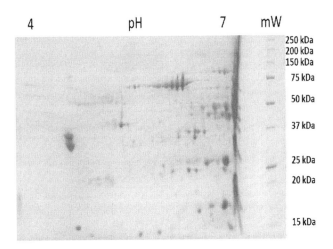

Figure 2. Representative 2DPAGE gel map of exudate in the pH range 4-7.

at day 1 compared to day 18 and 9 spots were more intense at day 18. Finally, 1 spot was more concentrated at day 6.

Conclusions

Our preliminary proteomic data from exudates of *Nero di Parma* dry cured hams are consistent with a differential release of specific proteins as a function of salting time. Peptide mass fingerprinting analysis are undergoing for the identification of proteins that are differentially extracted.

References

Paredi, G., Raboni, S., Bendixen, E., de Almeida, A.M. and Mozzarelli, A., 2012. 'Muscle to meat' molecular events and technological transformations: The proteomics insight. Journal of Proteomics 75: 4275-4289.

Paredi, G., Raboni, S., Dowle, A., Ashford, D., Thomas, J., Thomas-Oates, J., Saccani, G., Virgili, R. and Mozzarelli, A., 2013. The role of salt in dry cured ham processing characterized by LC-MS/MS-based proteomics. In: A. de Almeida, D. Eckersall, E. Bencurova, S. Dolinska, P. Mlynarcik, M. Vincova and M. Bhide (Eds.), Farm animal proteomics 2013, 2013/01/01. Wageningen Academic Publishers, pp. 274-277.

Pioselli, B., Paredi, G. and Mozzarelli, A., 2011. Proteomic analysis of pork meat in the production of cooked ham. Molecular BioSystems 7: 2252-2260.

Sabbioni, A., Beretti, V., Manini, R., Cervi, C. and Superchi, P., 2010. Application of different growth models to 'Nero di Parma' pigs, Proceedings of the 18th ASPA Congress, Palermo, June 9-12, 2009.

Can zymographic analysis of proteases activities provide new informations on 'foie gras' cooking losses?

Hervé Rémignon[1,2,3]*, Nathalie Marty-Gasset[1,2,3] and Sahar Awde[1,2,3]

[1]INRA, UMR1388 GENEPHYSE, 31326 Castanet-Tolosan, France; remignon@ensat.fr

[2]Université de Toulouse INP-ENSAT, UMR1388 Genephyse, 31326 Castanet-Tolosan, France

[3]Université de Toulouse INP-ENVT, UMR1388 Genephyse, 31076 Toulouse, France

Introduction

Fatty livers, also called 'foie gras' by gastronomes, are issued from the force-feeding of waterfowl, mainly Mule ducks (*Caïrina moschata* × *Anas platyrhynchos*). Foie gras is often commercialized and consumed cooked through sterilization or pasteurization. The cooking procedure of raw fatty liver results in lipids and water release from hepatic tissue due to its thermal denaturation. The extend of cooking losses is of great importance for processors because it affects their output and thus their profitability as well as it alters organoleptic and sensorial properties of the final product. It is well known that several factors can influence the degree of cooking loss in fatty liver, i.e. the genetic type of force-fed palmipedes (Salichon *et al.*, 1994), fatty liver's weight (Blum *et al.*, 1990) and its total lipids content (Nir and Nitzan, 1976) as well as rearing factors such as the duration of force-feeding or the age of birds. Technological parameters such as cooking temperature and duration have been also shown to alter cooking losses values (Rousselot-Pailley *et al.*, 1992). However, despite all the set of measures taken to control its variability, the cooking yield of fatty livers remains the major quality issue that faces commercial plants of fatty liver.

Based on what is already known concerning cellular structure alterations due to proteases activities during post-mortem storage of meat (Lonergan *et al.*, 2010), we hypothesized a possible similar role in the determinism of the value of cooking loss of Mule duck fatty liver. This hypothesize was tested thanks to a principal component analysis.

Material and methods

Animals

Fatty livers used in this experiment were issued from commercial flocks of male mule ducks reared until the age of 14 weeks according to standard practices.

Slaughter, fatty liver selection and sampling

At the end of the force-feeding period, the slaughtering procedure was conventionally conducted in a commercial plant. 20 min after bleeding, livers (500 to 550 g) were removed from carcasses and placed at +4 °C for 6 h. Selected livers were further sorted according to their melting rate by using the Melting Micro-Test Method (MMTM). Based on the result, the MMTM allowed distinguishing two extreme groups, i.e. M+ (high melting rate or high CL) and M– (low melting rate or low CL) of twenty livers each.

- Fatty liver cooking losses measurement: 100 g of fresh liver were used to determine CL after cooking in a water bath (85 °C) for one hour.
- Dry matter content: The dry matter content of fatty livers was determined by drying a mass of grounded liver in an oven at 105 °C for 24 hours.
- Lipid content: Total lipids content of raw samples was determined according to the Folch's method (1957).
- Protein extraction: Liver lysis was achieved using Fastprep-24 for 20 seconds followed by centrifugation 12,000×g at 4 °C for 15 min). For each extraction buffers, extractabilities (xtgelt and xtcalp for gelatine and casein degrading proteases extraction buffer respectively) were calculated by assuming a total protein content of 7% DM as reported by in a previous study.
- MMP-2 gelatine zymography: Matrix metalloproteinase 2 (MMP-2) gelatinolytic activity was detected according to Kizaki *et al.* (2008).
- Cathepsins gelatine zymography: Cathepsins activities were detected according to Afonso *et al.* (1997).
- Calpains casein zymography: Using the method of Raser *et al.* (1995), calpains activities were detected on 10% polyacrylamide gels containing 0.2% casein.

Image analysis

According to the procedure described by Bax *et al.* (2012), gels were scanned with an Image Scanner III using Image Master Platinum software (GE Healthcare, Uppsala, Sweden). Each liver sample was processed individually, and the value of main bands was calculated by using Image Master Platinum application.

Statistical analysis

For statistical analysis, the General Linear Model procedure of SAS software (SAS, 2011) was used with a one-way ANOVA to compare M– and M+ samples. In addition, a Principal Component Analysis (PCA) was performed with all samples and the biochemical and proteomic measured variables.

Results and discussion

As expected mean liver weight was very similar (520 g) in both M+ and M– ducks and is typical for this program of force-feeding. Thanks to this initial sorting, the weight of fatty livers will not explain cooking loss differences. The percentage of cooking loss (CL %) was 1.72 fold higher ($P<0.001$) in M+ than in M– samples. Livers from the M+ group presented higher dry matter and total lipids contents than those from the M– group (53.7 vs 61% DM respectively, $P<0.001$).

As reported in previous studies (Theron *et al.*, 2012), the more lipids in the liver, the higher the cooking losses (r=0.76, *P*<0.001).

The MMP gelatine protease assay permitted the detection of a 72 kDa proteolytic band corresponding to the MMP-2, while three proteolytic bands were detected for the cathepsins proteolytic assay and calpains proteolytic assays revealed two bands (Figure 1). Results concerning extracted proteases relative activities showed similar zymograms profiles in both M+ and M– samples but clearly exhibited higher values in M+ than in M– samples excepted for the relative activity of Ct-1. Finally, total extracted proteases activities appeared to be 1.4 fold (*P*<0.001) higher in M+ than in M– fatty livers. Higher proteolytic activities might induce higher degradation of cellular components because proteases remain active during post mortem storage. This hypothesis was previously reported by Theron *et al.* (2011 and 2013). Then, we can hypothesize that the rate of cooling of livers post-mortem could influence the cooking losses by modulating, more or less, the total activities of proteases: the more proteases are active (under the influence of temperature for example) during post mortem time, the more the structure of hepatocytes might be altered and the more their lipid content will be exudated during cooking. In the present study, the cooling rate was the same for both M+ and M– livers (same process, same liver weights) but proteases activities were originally higher in M+ than in M– samples.

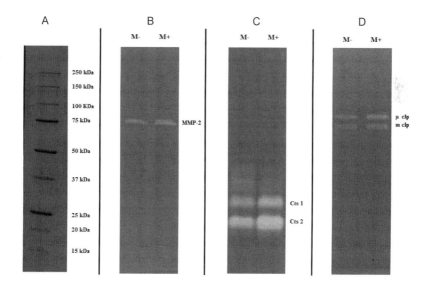

Figure 1. Molecular weight standards (A) and zymograms of the 72 kDa MMP-2 band (B), the 2 cathepsins bands (C) Cts-1 (30 kDa) & Cts-2 (25 kDa) and the μ- & m-calpains (μ-calp & m-calp) 80 kDa bands (D) analyzed by zymography in M– and M+ fatty liver samples.

The relationships between biochemical content, proteases activities and cooking losses are well illustrated by the principal component analysis (Figure 2) that clearly associates lipid and dry matter contents with proteases activities and cooking losses along the first principal component. The new linear combination of original variables given by the PCA allows a good distinction M+ and M– samples. It appears that M+ samples are mainly positively associated with lipid and DM contents as well as with proteases activities while it is the opposite for M– samples.

Conclusion

This study demonstrates that both lipid and DM contents as well as main hepatic proteases activities have a strong influence on cooking losses in fatty livers issued from force-fed Mule ducks but that lipid and DM contents remain the most determinants.

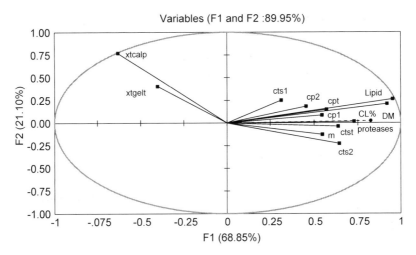

Figure 2. Plot of the first two principal component score vectors showing relationships between cooking losses and liver biochemical characteristics. m = matrix metalloprotease 2, cts1 = Cathepsin isoform 1, cts2 = cathepsin isoform 2, ctst = cts1+ cts2, cp1 = μ-calpain, cp2 = m-calpain, cpt = cp1 + cp2, proteases = m + cpt + ctst, Lipid = % lipid content, DM = % dry matter, xtgelt = extractability for gelatine degrading protease buffer, xtcalp = extractability for casein degrading buffer. ●: Added variable (CL%), ■: Active variables.

References

Afonso, S., Romagnano, L., Babiarz, B., 1997. The Expression and Function of Cystatin C and Cathepsin B and Cathepsin L During Mouse Embryo Implantation and Placentation. Development, 124, 3415-3425.

Bax, M. L., Chambon, C., Marty-Gasset, N., Remignon, H., Fernandez, X., Molette, C., 2012. Proteomic Profile Evolution During Steatosis Development in Ducks. Poultry Science, 91, 112-120.

Blum, J. C., Labie, C., Raynaud, P., 1990. Influence du poids et de la composition chimique du foie gras d'oie sur la fonte mesurée après stérilisation à 104 °C. Sciences des Aliments, 10, 543- 554.

Folch, J., Lees, M., Sloane Stanley, G. H. A., 1957. Simple Method for the Isolation and Purification of Total Lipides from Animal Tissues. Journal of Biological Chemistry, 226, 497-509.

Huff Lonergan, E., Zhang, W., Lonergan, S. M., 2010. Biochemistry of Postmortem Muscle – Lessons on Mechanisms of Meat Tenderization. Meat Science, 86, 184-195

Kizaki, K., Ushizawa, K., Takahashi, T., Yamada, O., Todoroki, J., Sato, T., Ito, A., Hashizume, K., 2008. Gelatinase (MMP-2 and -9) Expression Profiles During Gestation in the Bovine Endometrium. Reproductive Biology Endocrinology, 6, 66.

Nir, I., Nitsan, Z., 1976. Goose Fatty Liver Composition as Related to the Degree of Steatosis, Nutritional and Technological Treatments, and a Simplified Method for Quality Estimation. Annales de Zootechnie,25, 461-470.

Raser, K. J., Posner, A., Wang, K. K. W., 1995. Casein Zymography: A Method to Study μ-Calpain, M-Calpain, and Their Inhibitory Agents. Archives of Biochemistry and Biophysics, 319, 211-216.

Rousselot-Pailley, D., Guy, G., Sellier, N., Blum, J., 1992 Influence des conditions d'abattage et de réfrigération sur la qualité des foies gras d'oie. INRA Productions Animales, 5, 167-172.

Salichon, M., Guy, G., Rousselot, D., Blum, J., 1994. Composition of the 3 Types of Foie-Gras – Goose, Mule Duck and Muscovy Duck Foie-Gras. Annales de Zootechnie, 43, 213-220.

SAS, 2011 The Data Analysis for This Paper Was Generated Using SAS/STAT Software, Version 8 of the SAS System for Windows. Copyright © 2011, SAS Institute Inc. SAS and All Other SAS Institute Inc. Product or Service Names Are Registered Trademarks or Trademarks of SAS Institute Inc., Cary, NC, USA.

Theron, L., Cullere, M., Bouillier-Oudot, M., Manse, H., Dalle Zotte, A., Molette, C., Fernandez, X., Vitezica, Z. G., 2012. Modeling the Relationships Between Quality and Biochemical Composition of Fatty Liver in Mule Ducks. Journal of Animal Science, 90, 3312-3317.

Theron, L., Fernandez, X., Marty-Gasset, N., Chambon, C., Viala, D., Pichereaux, C., Rossignol, M., Astruc, T., Molette, C., 2013. Proteomic Analysis of Duck Fatty Liver During Post-mortem Storage Related to the Variability of Fat Loss During Cooking of 'Foie Gras'. Journal of. Agriculture and Food Chemistry, 61, 920-930.

Theron, L., Fernandez, X., Marty-Gasset, N., Pichereaux, C., Rossignol, M., Chambon, C., Viala, D., Astruc, T., Molette, C., 2011. Identification by Proteomic Analysis of Early Post-mortem Markers Involved in the Variability in Fat Loss During Cooking of Mule Duck 'Foie Gras'. Journal of Agriculture and Food Chemistry, 59, 12617-12628.

Peptidomics as a robust and reliable approach to discriminate between closely-related meat animal species

Alberto Massa[1], Enrique Sentandreu[1], Carlos Benito[2] and Miguel A. Sentandreu[1*]

[1]Instituto de Agroquímica y Tecnología de Alimentos (CSIC). Avenida Agustín Escardino, 7. 46980 Paterna (Valencia), Spain; ciesen@iata.csic.es

[2]Instituto de Gestión de la Innovación y del Conocimiento (CSIC-UPV). Camino de Vera s/n, 46022 Valencia. Spain

Introduction

There is an increasing demand by consumers for clear, reliable and detailed information about the foods they consume. This is especially relevant in processed foods, where ingredients cannot be distinguished by simple visual inspection. In this context, legislation must protect consumers against misdescription and fraud, practices that can be done by food producers or traders with the objective to increase the economic gain. The scandal occurring recently in Europe about the presence of undeclared horse meat in beef products illustrates this situation and highlights the importance to dispose of robust and reliable methods capable to unambiguously identify those species that are susceptible to be employed in fraudulent practices or as proof to certify the authenticity of the higher quality meats (Sentandreu and Sentandreu, 2014). In the case of meat products, there is a requirement to separately indicate and quantify the different meat species that are present in the food, what it is known as Quantitative Ingredient Declaration (QUID). In addition to this, other parts of carcass such as the liver, lung, heart or tongue, for example, cannot be considered as meat and need also to be separately indicated (Zukal and Kormendy, 2007).

Methods that have been traditionally employed in control laboratories to assess meat composition have mainly relied on immunoassays and DNA analysis. Even if they have remarkable advantages and performance, it is also true that they are not exempted from some important limitations, especially in the analysis of complex and/or highly processed foods. In the case of immunoassays, the lack of highly specific antibodies can promote the apparition of cross-reactions, something that becomes more probable when trying to differentiate between closely-related species such as the case of chicken (*Gallus gallus*) and turkey (*Meleagris gallopavo*) meat. In addition to this, food processing can greatly alter protein structure and consequently reduce its recognition by the antibody. Food processing can also negatively affect genetic analyses because DNA can undergo a remarkable degradation due to the use of aggressive conditions such as pH changes or thermal treatments, for example. This would increase the generation of shorter non-specific DNA fragments (Primrose *et al.*, 2010). Current advances in mass spectrometry applied to the analysis of proteins and peptides constitute an interesting and promising alternative to the aforementioned methods for the unambiguous identification of the different types of meats that can be present in meat products. The objective of the present work was to develop a peptidomic approach capable to discriminate between chicken and turkey, two closely-related meat species, through the identification and characterization of peptide biomarkers specific of each one of these farm animals.

Material and methods

One gram of either chicken (A) or turkey (B) meat were homogenized in 10 ml of 10 mM Tris buffer, pH 8.0, containing 6 M urea and 1 M thiourea by using a Polytron®. The homogenate was then centrifuged at 10,000×g for 20 min at 4 °C. From the obtained supernatant, an appropriate volume was taken in order to fractionate 2 mg of total protein by liquid isoelectric focusing in the pH range 4-7 using an Agilent 3100 OFFGEL fractionator. A total of 24 protein fractions were obtained using this technique. The protein distribution along these fractions was assessed by SDS-PAGE on 10% polyacrylamide gels. Twenty µl of fractions 2 and 3, which were found to contain myosin light chain 3 (MLC-3), were used for trypsin protein hydrolysis. After digestion, samples were completely dried using a speed-vac concentrator, then finally redissolved in 40 µl of 0.1% trifluoroacetic acid (TFA). Peptides generated by this way were subsequently analyzed by liquid chromatography coupled to electrospray ionization-tandem mass spectrometry (LC-MS/MS) using a LCQ Advantage ion trap instrument (Thermo Electron Corp.). Sequence identification of the biomarker peptides specific of either chicken or turkey species was done from the obtained MS/MS data using an in-house version of the Mascot search engine (www.matrixscience.com), together with the NCBInr protein database (www.ncbi.nlm.nih.gov/protein).

Results and discussion

In the methodology reported here, myosin light chain 3 (MLC-3) was selected as the target protein for the generation of peptide markers capable to unambiguously differentiate between chicken and turkey meats. As can be observed in Figure 1, isoelectric focusing of protein extracts allowed

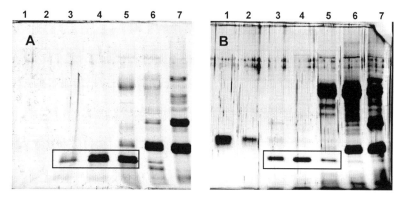

Figure 1. 10% SDS-PAGE corresponding to fractions 1 to 7 obtained after OFFGEL fractionation of either chicken (A) or turkey (B) meat extracts in the pH range 4-7. Bands corresponding to MLC-3 are those into the black-lined rectangle.

obtaining MLC-3 in fractions 3 and 4 separated from most of the rest of proteins, being an efficient enrichment step. MLC-3 also appears in fraction 5, though co-eluting with a higher number of contaminating proteins.

In order to evaluate the options to find species-specific peptides, it is necessary to compare the amino acid sequences of MLC-3 from chicken and turkey. The problem at this point is that turkey MLC-3 sequence is not currently available in the protein databases. However, it is available the sequence of turkey MLC-1. In the case of chicken, both MLC-1 and MLC-3 are available in the databases. This allowed us to know that most of the MLC-3 sequence (150 amino acids) is a part of MLC-1, which has a longer sequence (192 amino acids). As a consequence, we assayed a direct comparison between chicken MLC-3 and turkey MLC-1 as a way to find amino acid differences capable to generate marker peptides specific of each one of these two animal species. This comparison allowed us to observe that, despite the high homology, there would be three differing amino acid positions between chicken and turkey MLC-3, thus allowing the possibility for finding species-specific peptides.

From those fractions where MLC-3 appeared separated from the rest of muscle proteins (fractions 3 and 4), twenty microliters were taken to perform a trypsin digestion. The obtained peptides after this step were subsequently analyzed by LC-MS/MS. The identified peptides allowed confirming MLC-3 as to be the parent protein. Of the total identified peptides, two of them were specific of chicken and turkey, thus having potential for the unambiguous differentiation of these species in meat products. These two peptides, together with their main characteristics, are summarized in Table 1. The sequences would be homologous, being located in position 13-20 of the MLC-3 sequence.

Table 1. Peptides derived from the trypsin hydrolysis of MLC-3 characterized in the present work by tandem mass spectrometry (MS/MS) as specific of either chicken (Peptide A) or turkey (Peptide B) animal species. Differing amino acids between the two peptides appear in bold and underlined at the C-terminal position.

Peptide	Observed mass	Position	Sequence	Parent protein (protein entry name)[1]	Animal species
A	505.87 (2+)	13-20	EAFLLFD**R**	MLC-3 (gi|55584150)	*Gallus gallus*
B	492.05 (2+)	55-62[2]	EAFLLFD**K**	MLC-1 (gi|326922419)	*Meleagris gallopavo*

[1] Protein entry name in the NCBInr protein database.

[2] Position referred to turkey MLC-1, which is the protein whose sequence is available in the NCBInr protein database.

MS/MS data analysis allowed us to elucidate the amino acid sequence of these two peptides. In Figures 2 and 3 we can see the MS/MS spectra obtained for each one of them, together with the b and y ion series identified in each case. As it can be observed, the peptides have high sequence homology, only differing in the amino acid placed at C-terminal position. In the case of chicken (Figure 2), the C-terminal amino acid is arginine (R), whereas in the case of turkey we found lysine (K) at this position (Figure 3), being this the cause of the mass difference between these two peptides (Table 1)

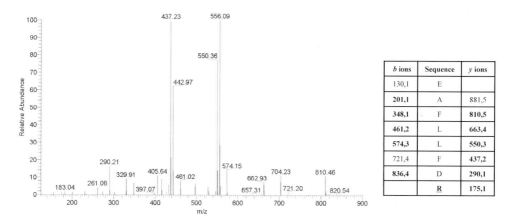

b ions	Sequence	y ions
130,1	E	
201,1	A	881,5
348,1	F	810,5
461,2	L	663,4
574,3	L	550,3
721,4	F	437,2
836,4	D	290,1
	R	175,1

Figure 2. MS/MS spectrum of peptide EAFLLFDR (Peptide A), generated from the trypsin hydrolysis of chicken (Gallus gallus) MLC-3. Identified b and y ions appear in bold in the table. Differing amino acid appears in bold and underlined at the C-terminal position.

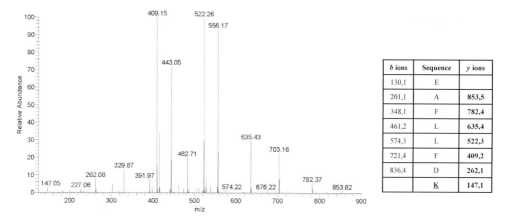

b ions	Sequence	y ions
130,1	E	
201,1	A	853,5
348,1	F	782,4
461,2	L	635,4
574,3	L	522,3
721,4	F	409,2
836,4	D	262,1
	K	147,1

Figure 3. MS/MS spectrum of peptide EAFLLFDK (Peptide B), generated from the trypsin hydrolysis of turkey (Meleagris gallopavo) MLC-3. Identified b and y ions appear in bold in the table. Differing amino acid appears in bold and underlined at the C-terminal position.

In this peptidomic approach, differentiation is made at the amino acid sequence level, thus its discriminating power being comparable to methods based on DNA analysis. However, since peptide sequences would be considerable more resistant to food processing than DNA sequences, this would allow for more reliable determinations especially in the case of highly processed foods where genetic material can undergo an important degradation (Sentandreu & Sentandreu 2014).

Conclusion

Results of the present work highlight the potential of Proteomics in the development of new control methods capable to accurately reveal the presence of the different types of meats that can be found in meat products. The proposed methodology, based on the identification of species-specific marker peptides, is robust, accurate and sensitive, allowing the possibility to discriminate between closely-related species such as the case of chicken and turkey meats. This peptidomic approach constitutes an interesting alternative to methods currently in use in revealing food composition and fraud control.

Acknowledgements

Financial support of projects AGL2012-32146 from the Spanish Ministry of Economy and Competitiveness and GV/2010/071 from Generalitat Valenciana are fully acknowledged.

References

Primrose,S., Woolfe,M., & Rollinson,S. 2010, Food forensics: methods for determining the authenticity of foodstuffs, *Trends in Food Science & Technology,* 21, 582-590.

Sentandreu,M.A. & Sentandreu,E. 2014, Authenticity of meat products: Tools against fraud, *Food Research International,* 60, 19-29.

Zukal,E. & Kormendy,L. 2007, On calculation of 'meat content' according to the quantitative ingredient declarations (QUID), *Journal of Food Engineering,* 78, 614-621.

Proteomic analysis of adipose tissue from peripartum high yielding dairy cows

Maya Zachut
Department of Ruminant Science, Volcani Center, P.O. Box 6, Bet Dagan, Israel;
mayak@volcani.agri.gov.il

Introduction

The modern high-yielding dairy cow faces a great metabolic challenge during the peripartum period: a shift from a non-lactating, late-gestational, and lipogenic period to a state of tremendous energy demand for milk production. This results in massive lipolysis of adipose tissue, as the cow is not able to consume sufficient nutrients to meet the energy requirements of the mammary gland (Bell and Bauman, 1997). In recent years the central role of adipose tissue as a regulator of metabolism in dairy cows was recognized, and many proteins expressed in this tissue are involved in the metabolic response to parturition and lactation (McNamara, 2012; Peinado *et al.*, 2012). Yet, little is known about the role of specific proteins in adipose tissue in these processes. Therefore, there is great potential in identifying novel proteins by proteomic analysis in the adipose tissue of high yielding dairy cows. To the best of our knowledge there is no data on global protein expression in adipose tissue from high yielding dairy cows. Therefore, the aim of this project was to identify differently expressed proteins in adipose tissues of high yielding dairy cows at late pregnancy and early lactation by mass spectrometry based proteomic analysis.

Materials and methods

In this study, eight high-yielding Israeli-Holstein multiparous dry cows (260 d in pregnancy) participated. Adipose tissue biopsies were collected at 17 d pre-partum and again at 3 to 5 d postpartum. The adipose samples were obtained from the subcutaneous fat pad over the gluteobiceps between the ischiatic tuber, tuber coxae and the base of the tail on the medial. The detailed method of adipose tissue biopsy was previously described (Zachut *et al.*, 2013). From each cow at each sampling time, four samples of about 40 mg of adipose tissue were collected and immediately frozen in -80 °C pending analysis. Proteins were analyzed by intensity based, label-free quantitative shotgun proteomics at The Israel National Center for Personalized Medicine Proteomics Unit, Weizmann Institute of Science (Rehovot, Israel). Briefly, proteins were extracted from the adipose tissue and then subjected to in-solution tryptic digestion. This was followed by nanoflow liquid chromatography coupled to high-resolution tandem mass spectrometry (nanoLC-MS/MS). Quantitative data was extracted using the Genedata Expressionist data analysis package and proteins identified using the Mascot search engine.

Results

In total, 586 proteins were detected in adipose tissues of high yielding dairy cows, and their expression was compared between prepartum and postpartum adipose samples. It was found that 103 proteins were significantly differentially expressed in adipose tissues prepartum as compared to postpartum adipose samples. From those, sixty-one proteins have known biological function and 42 proteins were un-characterized in cows. Analyzing the proteins that differed in expression pre- and postpartum revealed, for example, that the enzyme Succinyl-CoA ligase (ADP/GDP-forming) subunit alpha, mitochondrial, was 33 times more abundant in adipose tissues prepartum than postpartum ($P<0.05$). Also, the abundance of the lipogenic enzyme fatty acid synthase was 5 folds higher prepartum than postpartum ($P<0.02$). The enzyme 6-phosphogluconate dehydrogenase (decarboxylating) that is involved in the pentose phosphate cycle was 2.8 times more abundant in adipose tissue prepartum than postpartum ($P<0.02$). In addition, the expression of Galectin, a soluble protein with both intra- and extracellular functions was 3.7 times higher in prepartum adipose tissue relative to postpartum ($P<0.05$). In an opposite manner, the abundance of Ribosomal protein S14 was 3.5 times higher postpartum than prepartum ($P<0.001$). More details about the proteins mentioned above are shown in Table 1.

Table 1. List of several proteins that differ in expression pre- and postpartum in adipose tissue of high yielding dairy cows.

Protein description	Proteins with shared peptides	Protein MW	No. of proteins with shared peptides	No. of peptides	Unique peptides
Succinyl-CoA ligase (ADP/GDP-forming) subunit alpha, mitochondrial	tr|F1MZ38|F1MZ38_BOVIN	36,138	1	1	1
Fatty acid synthase	tr|F1N647|F1N647_BOVIN	274,264	1	50	50
6-phosphogluconate dehydrogenase, decarboxylating	tr|Q3ZCI4|Q3ZCI4_BOVIN	53,077	1	8	8
Galectin	tr|A6QLZ0|A6QLZ0_BOVIN	27,646	1	4	4
Ribosomal protein S14	tr|F1MR01|F1MR01_BOVIN, tr|Q3T076|Q3T076_BOVIN	16,776	2	1	0

Conclusions

This work is the first to demonstrate the global expression of proteins in adipose tissue of high yielding dairy cows. Preliminary analysis of the data revealed that several metabolic enzymes were more abundant prepartum, during late pregnancy when the adipose tissue is lipogenic. This data requires further bioinformatic analysis that will be performed in the near future. Nevertheless, this work indicates that the protein expression pattern in adipose tissues is considerably altered during the peripartum period, probably as part of the homeorhetic regulation to support lactation. Characterizing these proteins by proteomic analysis can improve our understanding of the processes that are involved in the metabolic adaptation to lactation in the adipose tissue of high yielding dairy cows.

References

Bell A. W., and D. E. Bauman. 1997. Adaptations of glucose metabolism during pregnancy and lactation. J. Mammary Gland Biol. Neoplasia. 2:265-78.

McNamara, J. P. 2012. Ruminant Nutrition Symposium: a systems approach to integrating genetics, nutrition, and metabolic efficiency in dairy cattle. J. Anim. Sci. 90:1846-54.

Peinado, J. R., M., Pardo, O. de la Rosa, and M. M. Malagón. 2012. Proteomic characterization of adipose tissue constituents, a necessary step for understanding adipose tissue complexity. Proteomics. 12:607-20.

Zachut, M., H. Honig, S. Striem, Y. Zick, S. Boura-Halfon, and U. Moallem. 2013. Periparturient dairy cows do not exhibit hepatic insulin resistance, yet adipose-specific insulin resistance occurs in cows prone to high weight loss. J. Dairy Sci. 96:5656-69.

Part IV
Advancing methodology for farm animal proteomics

Detection of whey fraction common proteins of human and goat colostrum by MALDI-TOF/TOF

Cansu Akin[1], Sébastien Planchon[2], Jenny Renaut[2], Ugur Sezerman[3] and Aysel Ozpinar[1*]

[1]Department of Medical Biochemistry, Acıbadem University, İçerenköy Mahallesi, Kayışdağı Caddesi No:32 Ataşehir, İstanbul, Turkey; aysel.ozpinar@acibadem.edu.tr

[2]Department Environment and Agrobiotechnologies, Centre de Recherche Public – Gabriel Lippmann, 41, rue du Brill 4422 Belvaux, Luxembourg

[3]Department of Molecular Biology, Genetics and Bioengineering, Sabancı University, Sabancı Üniversitesi, Orta Mahalle, Universite Caddesi No: 27 İstanbul, Turkey

Introduction

Human milk is an important source for the breastfed infants as it provides a digestible source of amino acids to infants and also has an immunological protection and assist in developmental functions as well. Therefore, the role of human milk in infant development is well established due to the presence of proteins that are found in different fractions of human milk, which are milk fat globule membrane (MFGM) proteins and skimmed colostrum proteins composed of whey proteins and caseins. One of the main difference observed in human milk when compared with other species' milk is that the whey fraction of the skim milk is the major protein source. Therefore, it is important to determine the proteomic profile of the whey fraction in order to better understand milk biogenesis and milk protein functions.

Aim

The aim of this study is to determine shared whey proteins in the skim fraction of human and goat colostrums via 2-DiGE (Differential Gel Electrophoresis) and MALDI-TOF/TOF mass spectrometric analysis in the field of contributing towards determining the protein contents of a human milk substitute.

Hypothesis

As the protein content of goat milk is more digestible and at the same time it is more tolerable than cow milk, it is expected that goat milk will more closely resemble human milk.

Methods

In order to assess the proteomic differences and/or similarities between human and goat colostrum, the samples from healthy mothers (n=6) and healthy goats (n=6), collected at early stages of lactation (i.e. 12-24 hours postpartum) were analyzed using DiGE electrophoresis and MALDI-TOF/ TOF mass spectrometry techniques. Master images constructed from both sample groups were compared to differentially expressed proteins, which were then be excised from the gels, trypsin digested and run on MALDI-TOF/TOF mass spectrometer for identification.

Results

According to the data obtained from MALDI-TOF/TOF mass spectrometry analysis, 422 overlapped spots were found on the selected master gel of human colostrum and goat colostrum. According to the 2 DiGE gels and MS/MS data, 260 different proteins were detected in both human and goat colostrum samples. As for the overlapped spots; immunoglobulin (n=25), casein (n=20), lactoferrin (n=8), lactoglobulin (n=4), albumin (n=9) lactotransferrin (n=6) and lactalbumin (n=4) originated proteins were mostly identified proteins. In addition, for the human colostrum lactoferrin and lactotransferrin related proteins were mainly highlighted and for the goat colostrum casein and lactoglobulin related proteins were mainly highlighted with the fold change of 50.

Conclusion

The results obtained from this study represent an important concept as goat's milk might be considered as an alternative substitute of human milk in the future. Although further studies are required to establish the proteomic profile differences seen in human and goat colostrums, this preliminary study underlies the importance of manufacturing high quality dairy products or infants formulas.

Regional brain neurotransmitter levels and proteomic approach: sex, halothane genotype and cognitive bias

Laura Arroyo[1]*, Anna Marco-Ramell[1], Raquel Peña[1], Daniel Valent[1], Antonio Velarde[3], Josefa Sabrià[2] and Anna Bassols[1]

[1]*Dept. Bioquímica, Universitat Autònoma de Barcelona, Spain; laura.arroyo@uab.cat*
[2]*Institute of Neurosciences, Universitat Autònoma de Barcelona, Spain*
[3]*IRTA, Monells, Girona, Spain*

Introduction

The development of methods for assessing the affective or emotional states is a crucial step in improving animal welfare. The 'cognitive bias', defined as a pattern of deviation in judgment in particular situations, is used as a marker (optimistic or pessimistic) for the effects of the affective state on cognitive processes (Douglas *et al.*, 2012).

Moreover, central nervous system involvement and its association with indoleamine and catecholamine functions have received major consideration in investigations of the etiology of porcine stress syndrome (Adeola *et al.*, 1993). Therefore the organization of the response to a stressful situation involves the activity of several areas of the limbic system through neurotransmitters synthesis (Mora *et al.*, 2012).

Objectives

The aim of this study was to design a proteomic approach in brain tissue to identify the protein expression map in these areas.

On the other hand, we determined the concentration of indoleamines (5-hydroxyindole-3-acetic acid (5-HIAA) and serotonin (5-HT)) and catecholamines (noradrenaline (NA), dopamine (DA), 3,4-dihydroxyphenylacetic acid (DOPAC) and homovanillic acid (HVA)) in the amygdala, hippocampus and prefrontal cortex of a group of slaughtered pigs classified according their emotional state, sex and halothane genotype.

Material and methods

The study was carried out on 48 hybrids Large White × Landrace pigs housed at Institut de Recerca i Tecnologia Agroalimentàries (IRTA)-Monells facilities. Animals were trained to learn to discriminate positive and negative spatial cues and classified according to their emotional state during the cognitive bias test, when an ambiguous cue was presented.

The brain was quickly removed and hippocampus, amygdala and prefrontal cortex were excised, dissected, frozen in liquid nitrogen and stored at -80 °C. Before analysis the samples were weighted and homogenized (1:10 w/v) in ice-cold homogenization buffer (0.250 mM $HClO_4$, 0.100 mM $Na_2S_2O_5$, 0.250 mM EDTA). After centrifugation, the supernatant was used to determine the concentration of noradrenaline (NA), dopamine (DA), 3,4-dihydroxyphenylacetic acid (DOPAC), homavanillic acid (HVA), 5-hydroxyindole-3-acetic acid (5-HIAA) and serotonin (5-HT) using a high-performance liquid chromatography (HPLC) with electrochemical detection.

Moreover, 70 mg brain tissue was mixed with 0.6 ml lysis buffer (30 mM Tris-HCl, 7 M urea, 2 M thiourea, 4% CHAPS, Protease Inhibitor Cocktail (Sigma), pH 8) and desalted. The isoelectrofocusing was performed with 100 µg prefrontal cortex protein on 7 cm pH 3-10 immobilized IPG strips (GE Healthcare). Then, proteins were separated by molecular weight in a 12% SDS-PAGE gel and stained with silver nitrate.

Results and discussion

Regional distribution of brain monoamines showed similar patterns to those described in the literature (Piekarzewska *et al.*, 2000) and our previous results (Table 1; UFAW Meeting, 2013).

Between sexes, halothane-carrier males showed lower concentration of dopamine and its metabolites in the amygdala, whereas halothane-carrier females showed higher concentrations of serotonin and its metabolites in the hippocampus (Table 2).

Table 1. Neurotransmitter mean concentrations (ng/g tissue) in the amygdala, prefrontal cortex (PFC) and hippocampus in female and male pigs.

	Amygdala		PFC		Hippocampus	
	Female	Male	Female	Male	Female	Male
NA	191.82	203.32	159.23	163.31	169.88[b]	193.85[a]
DOPAC	101.12	77.84	15.04	12.61	3.15	3.35
DA	473.57	418.43	20.29	20.60	27.62	28.21
HVA	353.94	341.63	67.63	54.30	31.65	28.17
5-HIAA	312.85	313.20	115.29	116.66	135.17	138.08
5-HT	1,155.38	1,072.81	318.60	334.75	342.08	350.61
Catecholamines	1,141.28	1,041.22	265.92	247.53	232.75	253.59
Indoleamines	1,468.23	1,386.02	428.37	451.42	477.25	488.69

[a] Significant differences between groups ($P<0.1$).

Table 2. Neurotransmitter mean concentrations (ng/g tissue) in the amygdala, prefrontal cortex (PFC) and hippocampus in halothane-carrier (Nn) and halothane-free (NN) pigs.

	Amygdala		PFC		Hippocampus	
	NN	Nn	NN	Nn	NN	Nn
NA	218.53[a]	181.51[a]	169.51	154.06	177.72	184.90
DOPAC	94.96	87.78	14.66	13.18	3.44	3.10
DA	479.00[a]	427.68[a]	19.78	20.94	27.57	28.21
HVA	348.46	348.20	63.55	59.25	31.42	28.69
5-HIAA	299.33	322.78	112.44	119.03	125.92[a]	145.80[a]
5-HT	1,075.42	1,149.59	328.41	325.26	336.76[a]	354.43[a]
Catecholamines	1,140.95[a]	1,065.29[a]	270.64	245.01	240.62	244.89
Indoleamines	1,374.75	1,472.37	440.85	439.11	462.68[b]	500.23[b]

[a] Significant differences between groups ($P<0.05$).
[b] Significant differences between groups ($P<0.1$).

When considering the effect of cognitive bias on neurotransmitter profile, females defined in a positive emotional state showed higher concentrations of homovanillic acid ($P<0.05$), a metabolite of dopamine, in the amygdala, probably related to the tendency to decrease in dopamine concentration found in these animals (Table 3). This suggests a relationship between motivation and cognition.

Table 3. Neurotransmitter mean concentrations (ng/g tissue) in the amygdala, prefrontal cortex (PFC) and hippocampus in pigs defined in negative or positive emotional state.

	Amygdala		PFC		Hippocampus	
	Negative	Positive	Negative	Positive	Negative	Positive
NA	180.59	200.20	145.49	163.91	154.38	186.23
DOPAC	88.57	91.21	12.87	14.07	2.98	3.30
DA	401.47	458.58	15.58	21.38	26.23	28.20
HVA	301.73a	357.95a	54.17	62.55	27.77	30.33
5-HIAA	290.83	317.44	109.96	117.09	130.92	137.56
5-HT	1,160.24	1,110.37	308.71	329.85	331.01	348.85
Catecholamines	972.35	1,121.71	228.10	262.32	211.36	248.32
Indoleamines	1,451.07	1,427.82	418.67	443.64	461.93	486.41

[a] Significant differences between groups ($P<0.05$).

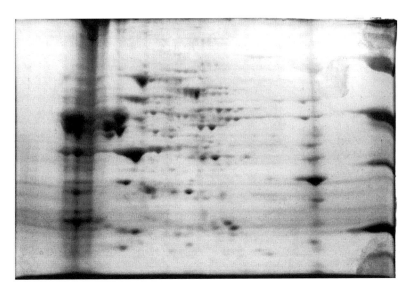

Figure 1. Silver-stained 2-DE gel from amygdala tissue.

Conclusions

We conclude that halothane genotype produces changes in neurotransmitter's concentration of slaughtered pigs. Furthermore, we also found differences in dopamine metabolism in the amygdala between female pigs defined as optimistic or pessimistic. On the other hand, we confirm proteomics as a possible tool for protein screening in brain tissues.

References

Adeola, O., Ball,. R.O., House, J.D. And O'Brien, P.J., 1993. Regional brain neurotransmitter concentrations in Stress-Susceptible Pigs. Journal of Animal Science 17:968-974.

Douglas, C., Bateson, M., Walsh, C., Bédué, A. and Edwards, S.A., 2012. Environmental enrichment induces optimistic cognitive biases in pigs. Applied Animal Behaviour Science 139: 65-73.

Mora, F., Segovia, G., del Arco, A., de Blas, M. and Garrido, P., 2012. Stress, neurotransmitters, corticosterone and body-brain integration. Brain Research 1476: 71-85.

Piekarzewska, A.B., Rosochacki, S.J. and Sender, G., 2000. The Effect of Acute Restraint Stress on Regional Brain Neurotransmitter Levels in Stress-Susceptible Pietrain Pigs. Journal of Veterinary Medicine Series A 47: 257-269.

Computational study of interaction of borrelial ospa with its receptors

Elena Bencurova[1]*[#], Dimitrios Vlachakis[3][#], Lucia Pulzova[1], Zuzana Flachbartova[1], Sophia Kossida[3] and Mangesh Bhide[1,2]

[1]*University of veterinary medicine and pharmacy in Košice, Komenskeho 73, Košice, Slovakia; bencurova.elena@gmail.com*

[2]*Institute of Neuroimmunology of Slovak Academy of Sciences, Dubravska cesta 9, Bratislava, Slovakia*

[3]*Bioinformatics & Medical Informatics Team, Biomedical Research Foundation, Academy of Athens, Athens 11527, Greece*

[#]*Authors contributed equally to this work*

Objectives

Lyme disease is the most common tick-borne zoonosis in Europe and North America. The causative agent *Borrelia burgdorferi* express on its cell surface numerous binding structures that are important during cycling between the vector and the mammalian host.

In this study, we focused on OspA protein, which is the major borrelial surface protein. OspA plays a significant role in the host-pathogen interaction and has multiple functions. OspA mediates adhesion of borreliae in tick gut endothelium via binding of tick receptor TROSPA (Pal *et al.*, 2004) and in the host it elicits strong humoral response. It was found that *Borrelia* is able to bind both human plasminogen and plasmin via OspA to disrupt the integrity of host cells and thus invade the tissues (Fuchs *et al.*, 1994). A series of experiments and reports have shown that OspA-CD40 interaction mediates adherence to endothelium and activate CD40 signaling pathway in endothelial cells (Pulzova *et al.*, 2011). Taken together, these findings clearly show multi-functional character of OspA.

In current work, we used the homology modeling to provide the three-dimensional structure of the OspA protein and its receptors: CD40, plasminogen and TROSPA. These models were docked for the further analysis of these important interactions and may help to predict potential binding domains of OspA.

Material and methods

Homology modeling

Homology modeling for OspA (*Borrelia bavariensis*, strain SKT 7.1) and its complexed receptors was performed using the Molecular Operating Environment (MOE) suite (Molecular Operating Environment (MOE)). Subsequent energy minimization was performed using the Gromacs-implemented, Charmm27 forcefield. The crystal structure of the outer surface protein A complexed with Fab from *B. burgdorferi* (PDB entry: 1OSP) was used as template for the modeling of OspA. Likewise the crystal structure of the complex of APC-Asef (PDB 3NMX) and crystal structure of the CD40 and CD154 complex (PBD 3QD6) was used as template structures for the modeling of the CD40 molecule. The crystal structure of the full-length type II human plasminogen (PBD 4DUR) was used for the modeling of plasminogen and the TROSPA, tick receptor, was modeled from the mitochondrial uncoupling protein 2 (PBD 2LCK).

Model optimization

Energy minimization was performed initially to remove the geometrical constrain from the top-ranking hits of the docking experiments, since the proteins were treated as rigid bodies so far. Protein complexes were subjected to an extensive energy minimization run using the Amber99 (Duan *et al.*, 2003) forcefield as it is implemented into the Gromacs MD suite, version 4.5.5, via the Gromita graphical interface, version 1.07 (Sellis *et al.*, 2009). An implicit Generalized Born (GB) solvation was chosen at this stage, in an effort to speed up the energy minimization process.

Molecular docking

In order to elucidate *in silico* the 3D structural conformation of CD40:OspA, plasminogen:OspA and TROSPA:OspA the docking suite ZDOCK, version 3.0 was used. RDOCK was utilized to refine and quickly evaluate the results obtained by ZDOCK (Chen *et al.*, 2003). RDOCK performs a fast minimization step to the ZDOCK molecular complex outputs and ranks them according to their re-calculated binding free energies.

Molecular dynamics

In order to further explore the interaction space and binding potential of each docking conformation, the molecular complexes were subjected to unrestrained molecular dynamics simulations using the Gromacs MD suite, version 4.5.5, and the Gromita graphical interface (Hess *et al.*, 2008). Molecular dynamics took place in a periodic environment, which was subsequently solvated with simple point charge water model using the truncated octahedron box extending to 7 Å from each molecule. Partial charges were applied and the molecular systems were neutralized with counter-ions as required. The temperature was set to 300 K and the step size was set to 1 femtoseconds. The total run of each molecular complex was fifty nanoseconds, using the NVT ensemble in a canonical environment. NVT stands for Number of atoms, Volume and Temperature that remain constant throughout the calculation. The results of molecular dynamics simulations were collected into a molecular trajectory database for further analysis.

Model evaluation

Prepared models were assessed within the Gromacs package by a residue packing quality function, which depends on the number of buried non-polar side chain groups and on hydrogen bonding. Further, the suite PROCHECK (Laskowski *et al.*, 1996) was apply to evaluate the quality of the produced models. Finally, the MOE suite was used to validate the 3D geometry of the models in terms of their Ramachandran plots, omega torsion profiles, phi/psi angles, planarity, C-beta torsion angles and rotamer strain energy profiles.

Results and discussion

In the present study, OspA and its receptors from host and vectors: CD40, TROSPA and plasminogen, were homology modeled and docked. OspA was modeled based on crystal structure of the complex of outer surface protein A:Fab (chain O), which gave a high sequence identity of 74% (Figure 1A, 1B and 1C). Even higher identity percentage was achieved in the search for a suitable template for plasminogen. That was the type II human plasminogen (chain A) with 79% identity (Figure 1A). The receptor CD40 was modeled by combining of templates CD40-CD154 chain R with 56% identities and Apc protein chain A, which shared 25% identity with the sequence of CD40 (Figure 1B). The most challenging homology modeling experiment was carried out for TROSPA. TROSPA is protein of a tick, which shares minimal identities with other proteins of known 3D structure. A series of four candidates were isolated for its homology modeling; PDB 2LCK (24% identity), 1O4U (26% identity), 1TUX and 3O2L (both 48% identity). The chain A of mitochondrial uncoupling protein 2 (PBD 2LCK) was used as template for this model (Figure 1C). After modeling, the models were *in silico* evaluated for their reliability. Initially, the models were structurally superimposed and compared to their templates, then they were re-evaluated with MOE and PROCHECK for their geometry.

The prepared models were then used in the docking experiment. Using the ZDOCK algorithm and interface, we constructed ligand-receptor models according to predicted or experimentally proved binding amino acids. For the plasminogen: OspA we did not find the interacting molecules, therefore the most prominent model was chosen. For the CD40:OspA and TROSPA: OspA a series of already identified amino acids was used to assess the protein interactions (Figlerowicz *et al.*, 2013; Mlynarcik *et al.*, 2013).

This study provides an insight to the host: pathogen interaction from the bioinformatics point of view. The results should be used for experimentally analysis of mutation or blocking of amino acids, which play role in pathogenesis and thus reduce the impact of the disease to host.

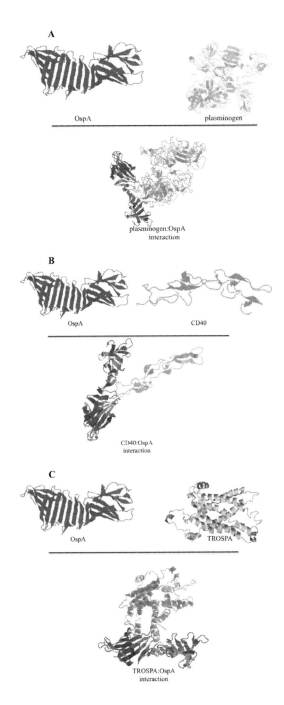

Figure 1. Homology modeling of OspA with its receptors. (A) homology modeling of OspA, plasminogen and the molecular docking conformation of plasminogen:OspA interaction (B) homology modeling of OspA, CD40 and the molecular docking conformation of CD40:OspA interaction, (C) homology modeling of OspA, TROSPA and the molecular docking conformation of TROSPA:OspA interaction.

Acknowledgements

Financial support was from APVV-0036-10, VEGA – 2/0121/11. STSM mobility was funded through COST-Action FA1002 Farm animal proteomics. E.B., L.P. and Z.F. are funded by ITMS 26220220185.

References

Chen, R., Li, L. and Weng, Z., 2003. ZDOCK: an initial-stage protein-docking algorithm. Proteins 52: 80-87.

Duan, Y., Wu, C., Chowdhury, S., Lee, M.C., Xiong, G., Zhang, W., Yang, R., Cieplak, P., Luo, R., Lee, T., Caldwell, J., Wang, J. and Kollman, P., 2003. A point-charge force field for molecular mechanics simulations of proteins based on condensed-phase quantum mechanical calculations. J Comput Chem 24: 1999-2012.

Figlerowicz, M., Urbanowicz, A., Lewandowski, D., Jodynis-Liebert, J. and Sadowski, C., 2013. Functional insights into recombinant TROSPA protein from Ixodes ricinus. PLoS One 8: e76848.

Fuchs, H., Wallich, R., Simon, M.M. and Kramer, M.D., 1994. The outer surface protein A of the spirochete Borrelia burgdorferi is a plasmin(ogen) receptor. Proc Natl Acad Sci U S A 91: 12594-12598.

Hess, B., Kutzner, C., van der Spoel, D. and Lindahl, E., 2008. GROMACS 4: Algorithms for highly efficient, load-balanced, and scalable molecular simulation. Journal of chemical theory and computation 4: 435-447.

Laskowski, R.A., Rullmannn, J.A., MacArthur, M.W., Kaptein, R. and Thornton, J.M., 1996. AQUA and PROCHECK-NMR: programs for checking the quality of protein structures solved by NMR. J Biomol NMR 8: 477-486.

Mlynarcik, P., Pulzova, L., Hresko, S., Bencurova, E., Dolinska, S., Kovac, A., Dominguez, M.A., Garrido, J.J. and Bhide, M.R., 2013. Identification of amino acid residues of OspA of Borrelia involved in binding to CD40 receptor, Farm animal proteomics 2013. Springer, pp. 107-111.

Molecular Operating Environment (MOE), b.C.C.G.I.

Pal, U., Li, X., Wang, T., Montgomery, R.R., Ramamoorthi, N., Desilva, A.M., Bao, F., Yang, X., Pypaert, M., Pradhan, D., Kantor, F.S., Telford, S., Anderson, J.F. and Fikrig, E., 2004. TROSPA, an Ixodes scapularis receptor for Borrelia burgdorferi. Cell 119: 457-468.

Pulzova, L., Kovac, A., Mucha, R., Mlynarcik, P., Bencurova, E., Madar, M., Novak, M. and Bhide, M., 2011. OspA-CD40 dyad: ligand-receptor interaction in the translocation of neuroinvasive Borrelia across the blood-brain barrier. Sci Rep 1: 86.

Sellis, D., Vlachakis, D. and Vlassi, M., 2009. Gromita: a fully integrated graphical user interface to gromacs 4. Bioinform Biol Insights 3: 99-102.

Recent advances in HRAM quantification: application to clinical assays

Bruno Domon

Luxembourg Clinical Proteomics Center, CRP-Santé, Luxembourg; bruno.domon@crp-sante.lu

The targeted analysis of protein biomarkers are routinely performed on triple quadrupole mass spectrometers operated in selected reaction monitoring (SRM) mode. However, the low resolution of quadrupole mass filters have limited selectivity, which is an issue for the analysis of complex samples, where the high background interferes with the signals of the analytes. Thus, hybrid mass spectrometers with high resolution and accurate mass (HRAM) capabilities overcome this limitation, and have opened new avenues in quantitative proteomics.

The targeted analyses of biological samples carried out using the parallel reaction monitoring (PRM) technique implemented on a quadrupole-orbitrap mass spectrometer have demonstrated a significant gain in selectivity, while the assignment of the fragment ions by accurate mass increased the confidence. The analyses in PRM mode showed better quantification performance for peptides present in low amount in bodily fluids. This translated in more consistent quantitative data for the different peptides of the same protein, and a clear discrimination between the control and patient samples.

Furthermore, optimal PRM performance was obtained through the control of the acquisition by monitoring the analytes in real-time and dynamically adjusting the parameters. The design of an instrument method called intelligent-PRM, yielded excellent results, even when applied to large sets of peptide in bodily fluids. Examples of application including the differentiation of isoforms will be discussed.

The sheep (*Ovis aries*) muscle proteome: decoding the mechanisms of tolerance to Seasonal Weight Loss using label free proteomics

Ana M. Ferreira[1,2#], Paolo Nanni[3#], Tanya Kilminster[4], Tim Scanlon[4], John Milton[5], Johan Greeff[4], Chris Oldham[4] and André M. Almeida[1*]

[1]*IICT – Instituto de Investigação Científica Tropical; CIISA – Centro Interdisciplinar de Investigação em Sanidade Animal, Lisboa, Portugal; Instituto de Tecnologia Química e Biológica / Universidade Nova de Lisboa and IBET – Instituto de Biologia Experimental e Tecnológica, Oeiras, Portugal; aalmeida@fmv.utl.pt*
[2]*ICAM – Instituto de Ciências Agrárias Mediterrânicas, Universidade de Évora, Évora, Portugal*
[3]*Functional Genomics Center Zurich, ETH/UZH, Zurich, Switzerland*
[4]*Department of Agriculture and Food Western Australia, Perth, WA, Australia*
[5]*University of Western Australia, Crawley, WA, Australia*
[#] *These authors contributed equally*

Introduction

Seasonal Weight Loss (SWL) is the most pressing issue in animal production in the tropics (Cardoso and Almeida, 2013). To counter SWL, farmers use supplementation with commercial feeds, expensive and difficult to implement (Cardoso & Almeida, 2013) or use animals that have a natural ability to withstand pasture scarcity, particularly fat tailed sheep breeds such as the Damara (Almeida, 2011) that have evolved in one of the most demanding ecosystems in the world, the fringes of the Kalahari desert in SW Africa.

Identification of markers of tolerance to SWL through proteomics will lead to increased stock productivity of relevant interest to animal production (Almeida *et al.*, 2010). The control of SWL has been the major goal of this research team for the last 20 years. We have been interested in the identification of molecular markers of tolerance to SWL, particularly at the muscle level, the tissue of capital economic importance in meat production.

Being muscle and meat essentially proteinaceous products, proteomics is hence of utmost importance to unravel the important physiological and biochemical mechanisms associated to all aspects of muscle and meat sciences (Paredi *et al.*, 2013), including SWL.

With this work, we aim to establish protein markers of tolerance to SWL in meat producing sheep. Samples from three breeds considered to have different levels of adaptation to nutritional stress were included: Merino (susceptible), Damara (tolerant) and Dorper (intermediate level). We will use a high-throughput label-free proteomics approach, that has seldom being used in this context.

Materials and methods

Muscle samples

Samples from each of these three breeds were colected during an animal assay, in which there were two nutritional groups per breed: underfed and control groups. Details of this experiment were recently published (Scanlon *et al.*, 2013). Briefly, the responses of the Damara, Dorper and Merino breeds to nutritional stress were compared during a 42 day trial. Seventy-two ram lambs, 24 from each breed, were randomly allocated to a growth (gaining 100 g/day) or a restricted diet (losing 100 g/day). Growth rates were determined (Scanlon *et al.*, 2013) and animals in the growth groups gained approximately 10% of their weight at the onset of the trial whilst the animals in the underfed groups lost approximately 15% of their weight at day 0. At the end of the 42 days, carcass characteristics were determined and the gastrocnemius muscle was sampled and frozen at -80 °C.

In the present study we compare the proteomes of the six experimental groups mentioned above: Damara Control, Damara Restricted, Dorper Control, Dorper Restricted, Merino Control and Merino Restricted.

Sample preparation

Nine samples per category, fifty-four samples in total (9×6 categories) were used for proteomics analysis. One hundred milligrams of muscle samples were added to 500 µl of ammonium bicarbonate (AMBIC) 50 mM, urea 8 M, Thiourea 2 M buffer, containing 1 tablet Complete Mini EDTA-free Cocktail and homogenized using an ultra-turrax instrument. The samples were then centrifuged for 6 minutes at 13000 rpm and the supernatant was removed and stored at -80 °C. For each sample a total of 15 µg of proteins were digested with trypsin using a Filter Aided Sample Preparation protocol (FASP) adapted from Wisniewski *et al.* (2009), and desalted previous to mass spectrometry analysis.

Mass spectrometry and data analysis

Peptides were loaded onto an in-house-made column (75 µm × 150 mm) packed with reverse-phase C18 material and analysed on an LTQ-Orbitrap Velos mass spectrometer. For each MS cycle, twenty Collision Induced Dissociation (CID) spectra were acquired in a data-dependent manner. Protein identification and label free quantification analysis were performed using Mascot (Matrixscience) and Progenesis software (Nonlinear Dynamics). A total of 685 proteins were identified and used for quantification (3.5% False Discovery Rate (FDR), minimum 2 peptides per protein). For protein quantification, 4 experimental designs were used: (A) Damara Control group vs Restricted group (9 vs 9 samples); (B) Merino Control group vs Restricted group 2 (9 vs 9 samples); (C) Dorper Control group vs Restricted group (9 vs 9 samples); (D) Control group vs Restricted group (27 vs 27 samples).

Results

The data analysis resulted in 685 proteins identified with ≥2 peptides (FDR: 3.5%). Looking at the comparisons between groups, we identified as differentially expressed the following proteins for each of the experimental designs (2 peptides, Anova<0.01, Fold change >2): (A) two proteins; (B) seven proteins; (C) five proteins; (D) four proteins. (Table 1.)

Discussion

In this study we report 685 proteins identified in sheep. To our knowledge, this is the first description of sheep muscle proteome. Moreover, we compare, within three sheep breeds, two experimental nutritional conditions: control group vs restricted groups. In a global comparison of control vs Restricted we identify four differently expressed proteins. Within each breed several proteins were found. Interestingly, in the Dorper breed, three of the five proteins are in the global comparison independently of breed. For the Merino two of the seven proteins are also in that list, and for Damara one of the two is also in that list. These results suggest that the four proteins we identified in D experimental design – apolipoprotein A-IV, immunoglobulin V lambda chain, ferritin heavy

Table 1. Differently expressed proteins found for the four experimental designs.

Experimental designs	Differentially expressed proteins
A	Immunoglobulin lambda light chain constant region segment 1 [Ovis aries]
	Immunoglobulin V lambda chain [Ovis aries]
B	PREDICTED: NADH dehydrogenase [ubiquinone] 1 alpha subcomplex subunit 6 [Ovis aries]
	PREDICTED: apolipoprotein A-IV [Ovis aries]
	Glutathione S-transferase [Ovis aries]
	Ferritin heavy polypeptide 1 [Ovis aries] [gi|410066835,gi|451327631]
	PREDICTED: collagen alpha-2(VI) chain [Ovis aries]
	PREDICTED: UPF0366 protein C11orf67 homolog [Ovis aries]
	PREDICTED: prolargin [Ovis aries]
C	Serpin peptidase inhibitor clade H member 1 [Ovis aries] [gi|410066852,gi|426245155]
	Ferritin heavy polypeptide 1 [Ovis aries] [gi|410066835,gi|451327631]
	Immunoglobulin V lambda chain [Ovis aries]
	PREDICTED: nebulin-related-anchoring protein [Ovis aries]
	PREDICTED: collagen alpha-1(XII) chain [Ovis aries]
D	PREDICTED: apolipoprotein A-IV [Ovis aries]
	Immunoglobulin V lambda chain [Ovis aries]
	Ferritin heavy polypeptide 1 [Ovis aries] [gi|410066835,gi|451327631]
	Serpin peptidase inhibitor clade H member 1 [Ovis aries] [gi|410066852,gi|426245155]

polypeptide 1, and serpin peptidase inhibitor clade H member 1 – might be interesting targets to continue our research looking for markers of SWL. Considering their function, these indeed seem interesting candidates. Apo A-IV is known to play a role in the regulation of food intake; immunoglobulin V lambda chain is involved in inflammation; ferritin heavy polypeptide 1 is the major intracellular iron storage protein in prokaryotes and eukaryotes. serpin peptidase inhibitor clade H member 1 is a heat shock protein that plays a role in collagen biosynthesis as a collagen-specific molecular chaperone.

Acknowledgements

This project was funded by PRIME-XS proposal #0000241 – The sheep (Ovis aries) muscle proteome: decoding the mechanisms of tolerance to Seasonal Weight Loss. Prime XS is funded by the 7th EU Framework program. AM Ferreira and AM Almeida are financed by Fundação para a Ciência e a Tecnologia, through grants SFRH/BPD/69655/2010 and SFRH/BPD/90916/2012 respectively.

References

Almeida, A.M. (2011). The Damara in the context of Southern Africa fat tailed sheep breeds. Tropical Animal Health and Production 43: 1427-1441.

Almeida, A.M., A. Campos, R. Francisco, S. van Harten, L.A. Cardoso and A.V. Coelho (2010). Proteomic investigation of the effects of weight loss in the gastrocnemius muscle of wild and New Zealand white rabbits via 2D-electrophoresis and MALDI-TOF MS. Animal Genetics 41: 260-272.

Cardoso LA, AM Almeida (2013). Seasonal weight loss – an assessment of losses and implications on animal welfare and production in the tropics: Southern Africa and Western Australia as case studies. Enhancing Animal welfare and farmer income through strategic animal feeding: 37-44. Edited by Harinder P.S. Makkar. FAO Animal Production and Health Paper No. 175. Food and Agricultural Organization of the United Nations (FAO), Rome.

Paredi G-L, M-A Sentandreu, A Mozzarelli, S Fadda, K Hollung, AM Almeida (2013). Muscle and meat: New horizons and applications for proteomics on a farm to fork perspective. Journal of Proteomics 88:58-82.

Scanlon TT, AM Almeida, A van Burgel, T Kilminster, JC Greeff, C Oldham (2013). Live weight parameters in Dorper, Damara and Australian Merino lambs subjected to restricted feeding. Small Ruminant Research 109: 101-106.

Wiśniewski JR, Zougman A, Nagaraj N, Mann M (2009). Universal sample preparation-method for proteome analysis. Nat Methods. 6:359-62.

Species determination of animal feed by QQQ mass spectrometry

Yue Tang, Janine Gielbert and Jim Hope
AHVLA Weybridge, New Haw, Surrey, KT15 3NB, United Kingdom

The EU TSE Roadmap II (2010-2015) has a strategic goal to review the current animal feed ban and consider appropriate revisions to feed legislation in line with the principles of proportionate and precautionary response. Since 1994, there has been a European-wide ban on the feeding of mammalian meat and bone meal to cattle, sheep and goats, and this ban was extended in 2001 to the use of processed animal proteins (PAP) in feed for any animals used for food (with exceptions such as the use of fish meal in pig and poultry feed). In Great Britain, the National Feed Audit monitors for the presence of prohibited ingredients of animal origin in feed and there is zero tolerance for breaches in the law. In considering the gradual relaxation of the feed ban, the EU TSE RoadMap II suggested a risk-based approach taking into account the availability of a reliable test to identify the species of trace amounts of MBM in feed, and its quantitation, with a view to introducing a tolerance level for feedstuffs. The current microscopic method is not suitable for the purpose of quantitation. In third countries (eg China), the polymerase chain reaction (PCR) amplification of species-specific DNA in feed is used to detect banned ruminant constituents but the allowed presence of ruminant milk in European animal feed and the poorer performance of PCR against heated samples precludes its use as a one-test screening and quantitation method in the member states of the European Union.

Currently, PAP excluded from the feed chain is used to produce fertiliser, compost or carburant for cement and its re-introduction in non-ruminant feed is viewed as highly beneficial as it would decrease EU (and GB)'s economic dependence on other sources of protein (eg soya bean). This measure would only be possible if validated analytical techniques for determining the species origin of PAP (and its quantitation) were available.

In this pilot study, we have investigated a targeted proteomics approach (selected reaction monitoring; SRM) for the speciation and quantitation of PAP. Our results provide initial quantitative data of the specificity and sensitivity of mass spectrometry (MS) for feed contamination with meat and bone meal (MBM) or mammalian processed animal protein (PAP) in pig and poultry feed or for second phase contamination of ruminant feed with these pig and poultry feeds. These data represent critical parameters for quantitative microbiological risk assessments (QMRAs) of PAP such as those produced by the European Food Safety Authority (EFSA) to underpin European Community (EC) policy on Feed ingredients (EFSA, 2010).

During our investigations, we encountered a lowering of sensitivity inversely proportional to the amount of sample applied to the mass spectrometric pre-analysis chromatography column. This matrix effect is a practical barrier to the routine application of this methodology and several approaches to solving this problem are being investigated. We hope to build on this work by varying sample preparation conditions, and use ion-exchange chromatographic columns to pre-purify trypsinised samples prior to MS/MS.

Differences between CH1641 scrapie and BSE prions identified by quantitative mass spectrometry

Adriana (Janine) Gielbert[1]*, Yue Tang[2], Maurice J. Sauer[1] and James Hope[2]
[1]Department of Specialist Scientific Support, Animal and Plant Health Agency-Weybridge, Addlestone, United Kingdom; janine.gielbert@ahvla.gsi.gov.uk
[2]Department of Virology and TSEs, Animal and Plant Health Agency-Weybridge, Addlestone, United Kingdom

Introduction

Transmissible Spongiform Encephalopathies (TSEs) are a class of transmissible diseases thought of as 'protein-only', caused by an infectious conformational change of a normal isoform of the mammalian prion protein (PrP^C) to a protease-resistant, pathogenic form called PrP^{Sc} and resulting in severe neurodegeneration inevitably leading to death. TSEs can occur in various species of farm animal and include scrapie in sheep, bovine spongiform encephalopathy (BSE) in cattle, and, in North America, chronic wasting disease in deer and elk. TSEs occur as many variants or strains. There is substantial evidence that variant CJD resulted from cross species transmission following consumption of BSE-contaminated food (Will *et al.*, 1996).

Differences between TSE strains can be demonstrated by biochemical analysis of PrP^{Sc}. The first step is usually limited proteolysis of PrP^{Sc} by proteinase K (PK), generating protease resistant fragments (PrP^{res}) of different sizes which depend on the type of disease. The different polypeptide fragments (theoretically) map to strain-type specific, conformational differences in PrP^{Sc}. PrP^{res} can then be analysed by differential Western blotting or discriminatory ELISA tests. However, many biologically distinct TSEs exhibit similar profiles and the low resolution of these tests does not easily permit sub-type discrimination.

CH1641 and CH1641-like scrapie are difficult to distinguish from experimental BSE in sheep using existing Western blot (WB) techniques, and scrapie with similar WB profiles have been identified over the years (Foster and Dickinson, 1988; Hope *et al.*, 1999; Vulin *et al.*, 2011). In recent years, progress has been made with biochemical methods that allow differentiation between CH1641 scrapie and BSE (Jeffrey *et al.*, 2006; Baron *et al.*, 2008; acobs *et al.*, 2011; Taema *et al.*, 2012). Using a multiplex immunofluorometric assay, Tang *et al.* were able to distinguish between classical scrapie, CH1641 scrapie, atypical scrapie and ovine BSE based on the ratio between the N-terminal antibodies 12B2 and 9A2, and the core antibody 94B4 (Tang *et al.*, 2012). Here, we have applied a mass spectrometry based method to determine the PK cleavage sites (N-terminal amino acid profiling, N-TAAP) (Gielbert *et al.*, 2009; Gielbert *et al.*, 2013) and core tryptic peptides to explore the nature of the differences between PrP^{res} prepared from either CH1641 scrapie or experimental BSE infected ovine brain tissue.

Method

Frozen brain tissue samples from sheep infected with TSEs were obtained from the Animal and Plant Health Agency (AHPA, formerly AHVLA) Biological Archive.

Samples are homogenized and treated with PK, followed by complete denaturation in 6 M GuHCl and reduced, alkylated and digested with trypsin as described previously (Gielbert *et al.*, 2009, 2013).

Analogues of peptides representing relevant sequences of ovine PrP (Figure 1) were custom synthesized (min. 98% purity; Peptide Protein Research Ltd.) and used, without further purification, for method optimization and as external calibration standards for quantification. An LC-MS/MS (mSRM) method was developed using an Agilent 6410 triple quadrupole mass spectrometer interfaced with a Chip Cube and Agilent 1200 nano-HPLC system (Agilent, UK). A total of twelve N-terminal PK cleavage sites and ten tryptic PrP peptides were quantified in each of the samples.

Results

Initially, N-TAAPs were obtained for SSBP/1 classical scrapie, CH1641 scrapie and ovine BSE under standard PK treatment conditions of 100 µg per millilitre 10% homogenate, incubating for 60 min at 50 °C (Figure 2). While the profile for classical scrapie is notably different from CH1641 scrapie and ovine BSE, no particular difference between the ovine BSE and CH1641 scrapie profiles was identified, a situation similar to past findings by WB (Foster and Dickinson, 1988; Hope *et al.*, 1999; Vulin *et al.*, 2011) and N-TAAP (Gielbert *et al.*, 2009).

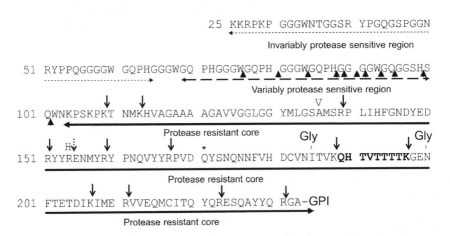

*Figure 1. Sequence and PK and tryptic cleavage sites for ovine PrP. Sequence is for ARQ genotype, with 136 and 154 polymorphisms as indicated and * denoting the 171 Q/H/R polymorphism. N-Terminal PK cleavage sites; Ü tryptic cleavage site. Bold sequence: tryptic peptide containing the 94B4 epitope.*

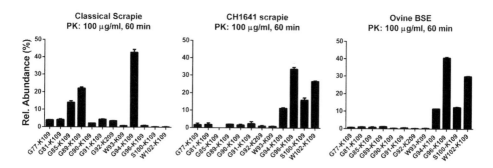

Figure 2. N-TAAP plots for three ovine TSEs: classical scrapie (VRQ/VRQ genotype), CH1641 scrapie (AHQ/AHQ) and ovine BSE (ARQ/ARQ). Data points (±SD) were obtained based on a single animal sample processed in duplicate.

To investigate whether any differences between CH1641 scrapie and BSE would be more evident under different PK conditions, we studied the breakdown of PrP^{Sc} and quantified both N-TAAP and core tryptic peptides for increasing PK stringency. The absolute abundance of either the sum of the N-TAAP peptides or the individual core peptides was relatively constant as a function of PK stringency for both classical and CH1641 scrapie. In contrast, the absolute abundance of BSE N-TAAP peptides was approximately ten-fold higher for incubation at 50 µg/ml PK and 15 min, then remained constant under the more stringent conditions. The BSE core peptide abundance was about three-fold higher for 50 µg/ml PK and 15 min incubation compared to 200 µg/ml and 90 min incubation (data not shown).

Abundances of the individual N-TAAP peptides were determined relative to the core tryptic peptide Q189-K197, which contains the 94B4 antibody epitope (Tang *et al.*, 2012), to compensate for variations in total PrP^{Sc} concentration between the homogenates. When the breakdown curves for CH1641 scrapie and ovine BSE were compared (Figure 3), marked differences were identified in the behaviour of both the N-terminal and core tryptic peptides. Looking at the N-TAAP peptides, there does not appear to be a notable overall change in their relative abundance for CH1641 scrapie, while for ovine BSE, the abundance of G96-K109 and G94-K109 appears to decrease with increasing stringency, while the abundance of W102-K109 increases.

To compare the individual C-terminal peptides for the different TSEs, their abundances were normalised to the concentration of Q189-K197 as for the N-TAAP peptides and, as they were fairly constant at any condition tested, averaged over the range of conditions. We found that the core peptides H114-R139, P140-R151, and Y152-R159 (CH1641) or E155-R159 (BSE) appeared to have a much lower relative abundance for CH1641 scrapie than for either classical scrapie or ovine BSE (Figure 4). To determine which differences were statistically significant, t-tests were carried out comparing each peptide between the TSEs (Table 1).

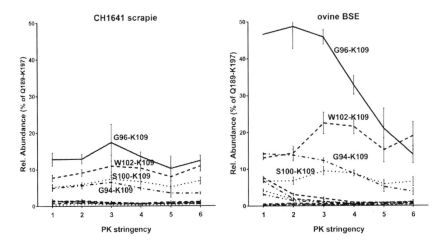

Figure 3. N-TAAP peptide breakdown plots as function of PK stringency, normalised to the abundance of Q189-K197 (the core tryptic peptide containing 94B4 epitope) at various levels of PK stringency (X-axes): 1=50 µg/ml, 15 min; 2=100 µg/ml, 15 min; 3=100 µg/ml, 30 min; 4=100 µg/ml, 60 min, 5=100 µg/ml, 90 min; 6=200 µg/ml, 90 min; all at 50 °C. Data points (±SD) were obtained from two different animal samples each processed in duplicate.

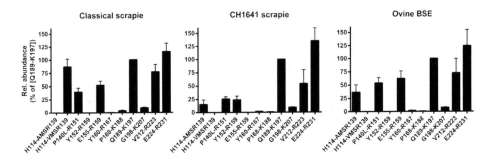

Figure 4. Core and C-terminal peptide abundance relative to the tryptic peptide containing the 94B4 epitope, Q189-K197, for three ovine TSEs: classical scrapie (VRQ/VRQ genotype), CH1641 scrapie (AHQ/AHQ) and ovine BSE (ARQ/ARQ). Data points (±SD) were obtained by averaging the results of two different animal samples each processed in duplicate, for each of the six PK stringency values given in Figure 3.

These results indicate that there are significant differences in relative abundance of core peptides across the entire H114-R159 core region, and therefore in the structures of PrP^{Sc} aggregates between ovine BSE and CH1641 scrapie. The most likely explanation is that CH1641 prions contain more fragments, not included in our assay, in this stretch, compared to BSE. In addition, the abundance of V212-R223 appears reduced for CH1641 scrapie. Although the existence of further fragments

Table 1. P-*values for t-test of significant difference in relative abundance of core tryptic peptides (normalised to the abundance of Q189-K197).*

	H114	P140	Y152/E155	Y160	P168	G198	V212	E224
Cl.Scr./CH1641	<0.0001	<0.0001	<0.0001	<0.0001	0.1368	0.3112	0.0004	0.0022
Cl.Scr./BSE	<0.0001	<0.0001	0.0031	<0.0001	<0.0001	<0.0001	0.4868	0.2127
CH1641/BSE	<0.0001	<0.0001	<0.0001	0.7595	<0.0001	<0.0001	0.0205	0.1857

of CH1641 prions has been previously reported (Baron *et al.*, 2008;Jacobs *et al.*, 2011;Tang *et al.*, 2012), we were surprised to find that these appear to distributed more widely across the protein sequence than previously thought. Nonetheless, this may account for the elusive nature of CH1641 PrP^res fragments encountered so far.

Acknowledgements

We are thankful to the Department for Environment, Food and Rural Affairs (Defra), UK, for funding this work (Grant SE2014). Many thanks to Mark Arnold for supporting the initial phase of this work (Grant SC0207, Modelling the PK digestion of the abnormal prion).

References

Baron, T., Bencsik, A., Vulin, J., Biacabe, A.G., Morignat, E., Verchere, J., and Betemps, D. (2008). A C-terminal protease-resistant prion fragment distinguishes ovine 'CH1641-like' scrapie from bovine classical and L-Type BSE in ovine transgenic mice. *PLoS Pathogens* 4, e1000137. http://dx.doi.org/10.1371/journal.ppat.1000137.

Foster, J.D., and Dickinson, A.G. (1988). The unusual properties of CH1641, a sheep-passaged isolate of scrapie. *The Veterinary Record* 123, 5-8.

Gielbert, A., Davis, L.A., Sayers, A.R., Hope, J., Gill, A.C., and Sauer, M.J. (2009). High-resolution differentiation of transmissible spongiform encephalopathy strains by quantitative N-terminal amino acid profiling (N-TAAP) of PK-digested abnormal prion protein. *Journal of Mass Spectrometry* 44, 384-396. http://dx.doi.org/10.1002/jms.1516.

Gielbert, A., Davis, L.A., Sayers, A.R., Tang, Y., Hope, J., and Sauer, M.J. (2013). Quantitative profiling of PrP peptides by high-performance liquid chromatography mass spectrometry to investigate the diversity of prions. *Analytical Biochemistry* 436, 36-44. http://dx.doi.org/10.1016/j.ab.2013.01.015.

Hope, J., Wood, S.C., Birkett, C.R., Chong, A., Bruce, M.E., Cairns, D., Goldmann, W., Hunter, N., and Bostock, C.J. (1999). Molecular analysis of ovine prion protein identifies similarities between BSE and an experimental isolate of natural scrapie, CH1641. *Journal of General Virology* 80 (Pt 1), 1-4.

Jacobs, J.G., Sauer, M., Van Keulen, L.J., Tang, Y., Bossers, A., and Langeveld, J.P. (2011). Differentiation of ruminant transmissible spongiform encephalopathy isolate types, including bovine spongiform encephalopathy and CH1641 scrapie. *Journal of General Virology* 92, 222-232. http://dx.doi.org/vir.0.026153-0 [pii]

Jeffrey, M., Gonzalez, L., Chong, A., Foster, J., Goldmann, W., Hunter, N., and Martin, S. (2006). Ovine infection with the agents of scrapie (CH1641 isolate) and bovine spongiform encephalopathy: immunochemical similarities can be resolved by immunohistochemistry. *Journal of Comparative Pathology* 134, 17-29.

Taema, M.M., Maddison, B.C., Thorne, L., Bishop, K., Owen, J., Hunter, N., Baker, C.A., Terry, L.A., and Gough, K.C. (2012). Differentiating ovine BSE from CH1641 scrapie by serial protein misfolding cyclic amplification. *Molecular Biotechnology* 51, 233-239. doi: 10.1007/s12033-011-9460-0.

Tang, Y., Gielbert, A., Jacobs, J.G., Baron, T., Andreoletti, O., Yokoyama, T., Langeveld, J.P., and Sauer, M.J. (2012). All major prion types recognised by a multiplex immunofluorometric assay for disease screening and confirmation in sheep. *J Immunology Methods* 380, 30-39. http://dx.doi.org/10.1016/j.jim.2012.03.004.

Vulin, J., Biacabe, A.G., Cazeau, G., Calavas, D., and Baron, T. (2011). Molecular typing of protease-resistant prion protein in transmissible spongiform encephalopathies of small ruminants, France, 2002-2009. *Emerging Infectious Diseases* 17, 55-63.

Will, R.G., Ironside, J.W., Zeidler, M., Cousens, S.N., Estibeiro, K., Alperovitch, A., Poser, S., Pocchiari, M., Hofman, A., and Smith, P.G. (1996). A new variant of Creutzfeldt-Jakob disease in the UK. *Lancet* 347, 921-925.

Proteomics data from ruminants easily investigated using ProteINSIDE

Nicolas Kaspric[1,2]*, Brigitte Picard[1,2], Matthieu Reichstadt[1,2], Jérémy Tournayre[1,2] and Muriel Bonnet[1,2]*

[1]INRA, UMR1213 Herbivores, 63122 Saint-Genès-Champanelle, France;
nicolas.kaspric@clermont.inra.fr; muriel.bonnet@clermont.inra.fr
[2]Clermont Université, VetAgro Sup, UMR1213 Herbivores, BP 10448, 63000, Clermont-Ferrand, France

Objectives

A main challenge for scientists working on the efficiency of ruminant production and the quality of their products (meat, milk) is to understand which genes and proteins control nutrient metabolism and partitioning between tissues, or which genes and proteins control tissues growth and physiology (Bonnet *et al.*, 2010). Their researches have produced huge amounts of genomic data requiring a lot of bioinformatics analyses to extract meaningful biological context for proteins or genes in ruminants. A strategy to increase the efficiency and the robustness of data mining from ruminant genomics data is to develop an online workflow that integrates several analysis steps in one package as it has been partly done in Human (BioMyn web service (Ramirez *et al.*, 2012)). Here, we present ProteINSIDE, an online workflow to analyse lists of proteins or genes from ruminant species by gathering biological information provided by an overview of data from myriad of databases, annotations according to Gene Ontology (GO), the prediction of tissue secretome and proteins interactions networks.

Methods

ProteINSIDE is an online tool, freely available at www.proteinside.org by using an internet browser. A flow chart (Figure 1) details the proposed basic or customizable analyses and the four main modules of the workflow. Whatever the type of analysis, the workflow uses data from the input file and runs default scripts (basic analysis) or scripts according to settings selected by the user (customs analysis). The four modules of analysis were developed as follow:

- 'Identifier mapping' module is an overview of available biological information thank to an assembly of reviewed biological data from UniProt and NCBI databases, and available on the 'ID resume' web page
- 'Gene Ontology' module queries QuickGO (Huntley *et al.*, 2009) database, with an option to unselect (basic analysis) or to select (custom analysis) electronic annotations. Then it ranks the over- represented GO terms by comparing the frequencies of a GO within a dataset and within the genome by a Fisher exact test. The most linked GO terms are viewed as a GOTree chart that is an ordered tree layout network provided by the custom analysis

- 'Secreted proteins' module searches for signal peptide on protein's fasta sequences thanks to SignalP (Petersen *et al.*, 2011) and TargetP (Emanuelsson *et al.*, 2000) algorithms. To support this prediction, ProteINSIDE selects GO terms related to secretion processes. The GO terms are also analysed to predict proteins that are secreted by processes that do not involve signal peptides (Nickel, 2003)
- 'Protein-protein interactions (PPi)' module searches for PPi recorded among the 26 PPi databases of Psicquic (Aranda *et al.*, 2011). ProteINSIDE only imports reviewed and curated PPi, and constructs PPi networks using an online cytoscape view (Lopes *et al.*, 2010) either between proteins of the dataset or between proteins of the dataset and outside of the dataset (in same species or using orthologous gene products)

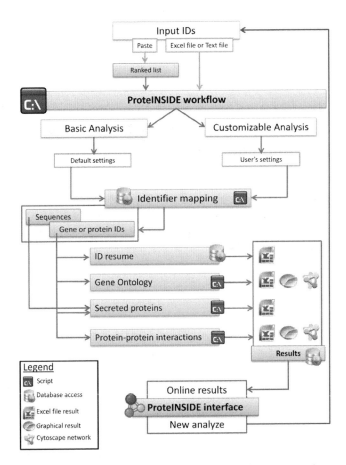

Figure 1. Flow chart of ProteINSIDE structure. The four modules for querying the available biological information, annotations according to the gene ontology, predictions of secreted proteins and protein-protein interactions are either all selected in the basic analysis or individually selected with specific settings in the custom analysis.

All results from ProteINSIDE's modules are viewed on the web page or can be downloaded. The output files are Excel file (.xls), Cytoscape file (.png or .pdf or .graphml or .xgmml), text or FASTA file (.txt or .fa) and pictures (.jpg or .png or .pdf).

ProteINSIDE inputs are proteins or genes identifiers (IDs) or names (e.g. ADIPO or ADIPO_ HUMAN or e.g. gi|62022275) or UniProt protein accession number (e.g. Q15848) from 6 species (bovine, ovine, caprine, human, rat, and murine). To test ProteINSIDE's performances, we used a dataset composed of 133 proteins: 34 proteins related to the glycolysis cycle, 11 proteins from the respiratory chain, 5 proteins from the tricarboxylic acid cycle, 79 hormones or secreted proteins and proteins with very specific functions unrelated to the others. We also included a duplicated protein among proteins of the glycolysis to verify its recognition by ProteINSIDE. We created this dataset on Bovine species, but the numbers of annotations and PPi weren't sufficient for a clear evaluation of the functionalities of ProteINSIDE. Then, we used the Bovine IDs for an analysis that use knowledge available in Human and ProteINSIDE automatically converted IDs from Bovine to Human using known orthologous gene products.

Results and discussion

Results from the 'Basic Analysis' shown that 133 proteins were recognized by ProteINSIDE, the protein in duplicate was identified and excluded from the analysis. Thus, 132 proteins were submitted to the analyses.

The ID Resume module of ProteINSIDE extracted and summarized biological information about each protein of the dataset. Results are available on the 'ID Resume' of the tool bar menu and are summarized in a table with: proteins and genes IDs, biological function, subcellular location, tissue specificity and chromosome location of the gene.

Among the 132 proteins submitted, ProteINSIDE annotates 123 proteins with 584 GO terms. GO terms are classed by the most significant p-value, by this way we retrieved the over-represented pathways expected for our dataset: GO terms relative to hormone activity and glycolytic process (that annotate respectively 33 and 27 proteins of the dataset). We have to note a lack of annotation for 12 proteins of the sample dataset, and a lack of annotation relative to glycolysis for 4 proteins (28 of the 33 expected proteins related to the glycolysis were annotated). This lack of annotations is related to our choice to use only GO terms that have been agreed by review curator in the 'Basic Analysis' (no annotation with IEA (Inferred by Electronic Annotation) evidence code). All proteins were annotated when we selected the use of IEA for GO in custom analysis.

ProteINSIDE predicts 85 as potentially secreted. Among the 85 proteins, 81 are confirmed by TargetP and 65 of them are both confirmed by subcellular location and GO terms annotation related to a secretory pathway. By merging shared results from the 3 analysis, we get 78 of the 79 proteins expected on our dataset. ProteINSIDE also predicted 31 proteins as potentially secreted by signal peptide-independent pathways.

The interaction researches between proteins of our sample dataset on 3 databases (BioGrid, IntAct, UniProt) have identified 29 PPi that involved 23 different proteins. As expected, PPi within our dataset linked proteins known to contribute to the pyruvate dehydrogenase complex, the complexes IV and I of the respiratory chain, and also some proteins linked to the glycolysis and the carbohydrate oxidation.

Similar results were obtained from bovine IDs submitted to analyses in bovine species with a custom analysis with settings that used electronic annotation and increased the numbers of PPi databases queried (BioGrid, IntAct, MINT, MatrixDB, STRING, Reactome, InnateDB, I2D and UniProt). This highlights that available data for ruminant species are poorly curated and mainly inferred from electronic annotations. Thus for ruminant IDs, we recommended to proceed to both a custom analysis in the ruminant species and to a basic analysis in a well annotated species (as Rat, Mouse or Human) and to compare the results.

Conclusion

In this work we present the performances of ProteINSIDE, a new powerful workflow which gather tools and public databases to retrieve biological information of genes or proteins lists from 6 species (Bovine, Ovine, Caprine, Human, Rat, and Murine). The presented web service has correctly identified a dataset of 133 proteins, has excluded a duplicate query and has retrieved biological information for each protein. According to our dataset, ProteINSIDE properly annotates the proteins related to the glycolysis, the proteins known as hormones, and the putatively secreted proteins. ProteINSIDE has revealed the most common pathways related to our dataset by creating networks from PPi within the dataset. Each result is easily accessible and downloadable. ProteINSIDE offers a great support to analyse a large quantity of data from genomic and proteomic studies. ProteINSIDE is also the unique web service that makes all of these analyses using ruminant IDs.

Acknowledgements

This work was supported by the regional council of Auvergne in France and APIS-GENE.

References

Aranda, B., Blankenburg, H., Kerrien, S., Brinkman, F.S., Ceol, A., Chautard, E., Dana, J.M., De Las Rivas, J., Dumousseau, M., Galeota, E., Gaulton, A., Goll, J., Hancock, R.E., Isserlin, R., Jimenez, R.C., Kerssemakers, J., Khadake, J., Lynn, D.J., Michaut, M., O'Kelly, G., Ono, K., Orchard, S., Prieto, C., Razick, S., Rigina, O., Salwinski, L., Simonovic, M., Velankar, S., Winter, A., Wu, G., Bader, G.D., Cesareni, G., Donaldson, I.M., Eisenberg, D., Kleywegt, G.J., Overington, J., Ricard-Blum, S., Tyers, M., Albrecht, M. and Hermjakob, H., 2011. PSICQUIC and PSISCORE: accessing and scoring molecular interactions. Nat Methods 8: 528-529.
Bonnet, M., Cassar-Malek, I., Chilliard, Y. and Picard, B., 2010. Ontogenesis of muscle and adipose tissues and their interactions in ruminants and other species. Animal 4: 1093-1109.

Emanuelsson, O., Nielsen, H., Brunak, S. and von Heijne, G., 2000. Predicting subcellular localization of proteins based on their N-terminal amino acid sequence. J Mol Biol 300: 1005-1016.

Huntley, R.P., Binns, D., Dimmer, E., Barrell, D., O'Donovan, C. and Apweiler, R., 2009. QuickGO: a user tutorial for the web-based Gene Ontology browser. Database (Oxford) 2009: bap010.

Lopes, C.T., Franz, M., Kazi, F., Donaldson, S.L., Morris, Q. and Bader, G.D., 2010. Cytoscape Web: an interactive web-based network browser. Bioinformatics 26: 2347-2348.

Nickel, W., 2003. The mystery of nonclassical protein secretion. A current view on cargo proteins and potential export routes. European Journal of Biochemistry 270: 2109-2119.

Petersen, T.N., Brunak, S., von Heijne, G. and Nielsen, H., 2011. SignalP 4.0: discriminating signal peptides from transmembrane regions. Nat Methods 8: 785-786.

Ramirez, F., Lawyer, G. and Albrecht, M., 2012. Novel search method for the discovery of functional relationships. Bioinformatics 28: 269-276.

Acknowledgements

The Organizing Committee of the Farm Animal Proteomics 2014 meeting acknowledges the support of the entities and individuals that made possible this publication and event.

Support by

 UNIVERSITÀ DEGLI STUDI DI MILANO
DIPARTIMENTO DI SCIENZE VETERINARIE E SANITÀ PUBBLICA

EUROPEAN COOPERATION IN SCIENCE AND TECHNOLOGY

Under the patronage of

RegioneLombardia

ITALIA
EXPO MILANO 2015

Milano

Comune di Milano

Printed in the United States
by Baker & Taylor Publisher Services